THE SAND DOLLAR
and the SLIDE RULE

The SAND DOLLAR

and the SLIDE RULE

Drawing Blueprints from Nature

DELTA WILLIS

ADDISON-WESLEY PUBLISHING COMPANY, INC.
Reading, Massachusetts • Menlo Park, California
New York • Don Mills, Ontario • Harlow, England
Amsterdam • Bonn • Sydney • Singapore • Tokyo
Madrid • San Juan • Paris • Seoul • Milan
Mexico City • Taipei

All acknowledgments for permission to reprint previously published material can be found on pages 233–234.

Many of the designations used by manufacturers and sellers to distinguish their products are claimed as trademarks. Where those designations appear in this book and Addison-Wesley was aware of a trademark claim, the designations have been printed in initial capital letters.

Library of Congress Cataloging-in-Publication Data

Willis, Delta.
    The sand dollar and the slide rule : drawing blueprints from nature / Delta Willis.
        p.   cm.
    Includes bibliographical references and index.
    ISBN 0-201-63275-6
    ISBN 0-201-48831-0 (pbk.)
    1. Morphology.   2. Biomechanics.   3. Nature (Aesthetics)
I. Title.
    QH351.W55   1995
    574.4—dc20   94-37157
                                                                                    CIP

Cover design by Suzanne Heiser
Text design by Joyce C. Weston
Set in 11-point Janson by G&S Typesetters, Inc., Austin, TX

1 2 3 4 5 6 7 8 9-MA-0099989796
First printing, December 1994
First paperback printing, June 1996

To Bob, the civil engineer
who would rather fly

CONTENTS

AUTHOR'S NOTE
THE LEGEND OF
THE SAND DOLLAR

The sand dollar that volunteered for the title sits on a bookshelf nearby as I write. It has superficial beauty to recommend it, as if an artist had etched a blossom with five petals onto thin, delicate sandstone. The dead form assumes what the French call the malice of objects, when something coveted has vanished and you fear imminent blindness because the thing was "just here." There is nothing left in the interior to explain the surface pattern; the integrity this former sea urchin knew has been lost, and the only thing it can produce is sand. Every time I pick it up to study a detail, a few grains slip out, putting grit in various research papers, my notes, and ideas that seem too smooth. This particular dollar has continued to yield since I found it on the Kenya coast in March of '91, and the manner in which it imparts clues serves as a reminder of the way science works. A mere three and a half inches in diameter, and less than a half an inch thick at its apex, it yields a volume scarcely suggested by its dimensions. It was still yielding when I wrote the last page of this book. The study of organic forms is not a static business.

This book explores the dynamics of organic forms, and how blueprints drawn from nature can improve the human condition. That we can peer inside forms such as this, and grasp nature from an engineering point of view, is a bonus, yet the critical difference between most human designs and organic forms remains the vitality of response. It was discovering this critical difference that allowed the Wright Brothers to succeed where others failed; they needed wings that could respond to shifts in wind currents; a turkey vulture provided demonstrations. Simply discovering the form of a blueprint is only the first step. Recreating the mechanics is more complex, and often defies our notion of stability, as unaccountably rigid as our notion of good engineering, firm and uncompromising. If by some fiat I were forced to hone my stance to a single sentence, it would be this Kikuyu saying: The truth is like a lizard's tail. You might seize it in your hand, meanwhile, the lizard moves on, creating a whole new tail.

I was drawn to this subject after several years of photographing African trees. Why do some branches weep, some rise like a *Candelabra euphorbia*, and others (*Euphorbia robecchii*) have such neat symmetry? I was hardly the first writer mesmerized by these forms. In *The Green Hills of Africa*, Ernest Hemingway wrote of the neat line of acacias "that runs parallel to the horizon, a pleasing arrangement to the human eye." This is known as the browse line, since it's as high as giraffe can reach to feed. On a safari in Tanzania's Selous, novelist Tom Robbins noticed "trees that look like Fifties haircuts, their foliage barbered into Sha Na Na flattops." These flat-topped acacias employ a broad crown for maximum exposure to the sun, the opposite tact of the African baobab, with branches that resemble roots. Legend has it that the baobab was planted upsidedown by an angry god. Legends and myths greeted most of my inquiries. There didn't seem to be a great deal of handy data on the way wild trees work. "The form of trees is perhaps their most distinctive feature as a biological group," wrote P.B. Tomlinson, based at the Harvard Forest; "yet we have only recently understood quite elementary features of their construction."

One of their elementary features is fractals, the bifurcation and repetition that occurs in tributaries and our own lungs. This geometrical pattern is testimony to a system of distribution that maximizes flow in economic ways. Can it be duplicated in human designs? It already occurs in some patterns of highways that fork from large arteries to small, and in the growth patterns of city streets. "Everything made by human hands and most things conceived by the human mind have their prototype in nature," wrote Andreas Feininger, the *LIFE* photographer and former architect, and looking at nature from an engineering point of view has already led to considerable profits. To gain new insights on why things take the shape they do, several teams of scientists and architects have formed what I call Think Again Tanks. They not only look for lessons in nature to improve upon existing designs, but many return to the history of human enterprise, such as the remarkable work of Scottish biologist D'Arcy Thompson, whose 1917 book *On Growth and Form* presented organic patterns as diagrams of forces. Gravity is an obvious force in the shape of our own bones; trees also respond to the loads they must carry, yet viscosity and surface tension dictate the boundaries of the minute.

Many experts see these blueprints not just as sources for profit, but models for thinking. Pliability is the best idea we could possibly steal from the organic. The flexibility demonstrated by natural forms is not based on drift and whim but inner strength. They bend with the wind, go with the flow, adapt, respond, and perhaps most important of all, they grow. In all cases, the mantra that surfaces is the ability to change. What sounds philosophical can be measured with some precision, by formulas for fluid dynamics, thermodynamics, aerodynamics, elasticity. Suddenly a form that merely appears elegant has numbers to explain its economy. That said, this book does not require a calculator, or even a vague recall of a hypotenuse.

The slide rule is used to represent a leap of magnitude by the powers of ten, and the advantages and disadvantages of tools. Do some of our tools—computers, slide rules, calculators—impair our skills of observation? By imposing formulas on a complex landscape, what details do we discard?

Harmony and symmetry in the universe have such a powerful appeal that scientists find themselves drawn to the elegant simplicity of math, and geometric patterns that distill and contain. If math can measure harmony, we reason, numbers should warn us of potential disasters in the greater flux. William Faulkner described the devastating floods of the Mississippi as an inevitability understood by locals: the river was "like a mule that worked for you for ten years for the privilege of kicking you once." There are moments when the earth itself seems to harbor a grudge, as it did during a 1989 earthquake in Oakland, California, when a freeway collapsed only along the section traversing landfill rather than natural sediments. Drawing blueprints from nature is not confined to isolated systems; it was D'Arcy Thompson's central thesis that forms respond to the physical forces of their environment as part of the greater flux.

There was a time when European travelers shielded their eyes from jagged mountain peaks, as nature in the raw was ugly and they wanted no part of it. Orchids and roses they might cultivate in their gardens were a different proposition; coffee and tea plantations, apple orchards, and neat vineyards fit in with domesticated pleasures. Control was paramount. Civilization was intent on taming the unruly and savage forces. Things should be orderly. Geometry began to lace the uneven terrain of Europe as it now does the bulk of every continent but Antarctica, Greenland, and parts of Africa. Right angles were convened with certain passion. They marked countries and counties, city blocks, buildings and bricks, pastures and fields, coffins and calling cards. It was unnatural. For all of the dazzling patterns in nature, there are precious few right angles. You can raise your arm to the side and create one. Tree branches come close, but like the arm pit, embrace a slight curve for support.

Many of the forms in this account are beautifully symmetrical, but for every law in nature there seems to be an outlaw, and for every basic blueprint, a multitude of variations, some maligned as miscellanea. I seized upon the sand dollar because it was described as irregular when compared to its cousins with their round dome. Such extremes can reflect a pathway to what physicist Peter Allen calls a "possibility space," a potential form that does not exist. Even the pathway for modifications can be perilous, as Buckminster Fuller learned with his Dymaxion car, and some of the more simple blueprints in nature may have been misunderstood, as paleontologist Dolf Seilacher suggests with early life forms found in Australia.

All organic forms do not qualify for copyright. The bottom line is maximum performance with a minimum use of energy. A hummingbird, for example, uses an enormous amount of fuel for hovering, and is often compared to a poorly designed helicopter. In the quest for so-called optimal forms, the wheel verges on embarrassing, saved only by the spokes of a bicycle. A study of the energy cost of locomotion (calories consumed × distance × time) reveals that a human on a bicycle is the second most efficient form of locomotion. First place goes to a trout. The advantages of fish include a tail that can move a volume of water several times their own body weight, and a design with minimum drag.

A sand dollar employs the same principles of fluid dynamics to maintain stability, its slender dome shaped like a flying saucer. It has various mechanisms for balancing lift and drag. This abundant sand that spills on my desk served as ballast, and the one I found in Kenya apparently loaded up with all the ballast it could possibly hold, a casualty of a fierce storm at high tide. Two keyholes marked this echinoid from the Indian Ocean as belonging to the species *Echinodermata biserforatus;* in others from the east coast of the U.S., these holes can number five. These holes fill with mud or clay to serve as lift spoilers, an example of Form following Function.

Sea urchins, flat and domed, have been studied since Aristotle, who applied the name *echinos,* or hedgehog, in reference to their spines. Nearly a century after Darwin published *On the Origin of Species,* the preeminent surveyor of invertebrates at the American Museum of Natural History, Libbie Hyman, saluted the sand dollar's phylum, *Echinodermata,* "as a noble group especially designed to puzzle the zoologist."

For a long time, organic forms were studied simply to classify species by descent. Their parts were used to link them to others; the sand dollar, for example, is related to the starfish by their common radiation of pentagons. When you flip over an ex-urchin, on the bottom is an outline resembling a poinsettia, five petals again, like the five petals on the surface. It never occurred to me to relate these configurations of five to my own fingers. Such a connection of patterns seemed dubious, as suspect as astrology, or the flipped perspective of a beachcomber who exclaims, "Hey, look! All these bivalves have two parts!" Linking patterns may be the only way to discover similar forces at work, but it can be precarious terrain, whether you use math or myth. "The Legend of the Sand Dollar" popped up on a rack of postcards in a pharmacy in Beaufort, North Carolina.

Here's a lovely little story
That so many men will tell,
Of the life and death of Jesus
Etched upon this lonely shell.

If you look at it real closely,
You will find an image here
Of four nails and then another
from a Roman's sharpened spear.

One side shows the Easter Lily
With its center as a star,
that shined brightly for the shepherds
As they traveled from afar.

And the Yuletide poinsettia
Painted on the other side
Tells us Christ was born on Christmas,
Wore our cross until he died.

If you break the center open,
You will find the sign of peace
Five white doves in gleaming beauty.
Will its wonders never cease!

So you see the simple story:
Jesus lived for you and me.
To carry on his work on earth,
To love humanity.

The five white doves are actually part of a chandelier of tendon and teeth known as Aristotle's lantern, an instrument of such utility that it brings to mind one word: Patent. Yet the pyramids and pentagons that shape it have already been employed. What remains as a potential is the

The pattern of five on the surface of the sand dollar is a result of five tube feet in the interior, and the poinsettia-like pattern on the bottom (as seen on the cover of this book) are food grooves that process vast amounts of sand moved by water.

novel rearrangement of these parts, with fluid and elastic connections, as happened with one of the most beautiful achievements of humankind, a concord of numbers and time. Music may be the human invention that most resembles evolution, because so many variations arose from just a few basic elements.

Among the trees of Africa, there is a scrub acacia known as the Whistling Thorn. At night when the wind blows, these trees sing an eerie song. The music derives from delicate, black spheres, or galls, that serve as a home for ants, which leave a little hole when they depart their domes, creating wind instruments. All trees apparently have another, individual harmony, a singular frequency that denotes their ability to bend with the wind.

THE SAND DOLLAR
and the SLIDE RULE

Two acacias at sunset on the Chyulu Hills in Kenya, where smaller acacias on the surrounding savannah are called Whistling Thorns. Mount Kilimanjaro is on the left in the distant clouds.

1. The Think Again Tank

*He ran his finger over a heavy-grained oak surface, and to him it
was an exquisite pleasure, vibrating in his veins like music. . . .
What he loved so much in plant morphology was, that, given a
fixed mathematical basis, the final evolution was so incalculable.*
— D.H. LAWRENCE, *Mr. Noon*

A HEFTY piece of tree limb was passed around a lecture hall near
Stuttgart. It traveled through the hands of architects, engineers,
paleontologists, and the occasional philosopher. The speaker was a
physicist, Claus Mattheck, who wore shaded wire rims, a dark brown
leather jacket, and a haircut distinctly medieval Beatle.

Mattheck works at the Nuclear Research Center in nearby Karlsruhe,
Germany, and is known to take long walks in the Palatinate Forest, where
he studies what he calls the "body language" of trees. He uses his data to
refine industrial designs: car parts, machine parts, even spare parts for the
human body. Trees, he likes to say, are the "design teachers of us."

Mattheck's audience was gathered for an international conference on
Natural Structures, a rare dialogue among disciplines begun by German
architect Frei Otto. Otto is best known for the undulating rooftops over
the 1972 Munich Olympic stadia. Some elements of that design were
inspired by the work of spiders; others resulted from playing with soap
bubbles, contorting the film membrane into a variety of forms with a
malleable wire. The spiders demonstrated lightweight strength; the soap
film, a minimum use of materials.

For four days, from eight in the morning until six in the evening, people drew upon blueprints from nature. A butterfly's proboscis was compared to the shaft of an oil drill. The pleats of a poppy were used to demonstrate how folding patterns strengthen even the most delicate material. The dome of a sea urchin featured as a model for vaulted inspirations: St. Mark's Cathedral in Venice, the Houston Astrodome, an igloo, a Volkswagen beetle, and, in a circuitous tribute ("Next slide, please"), the human skull. The Sydney Opera House was described as terrible from an engineering point of view because the pointed shells were incomplete domes and required edge beams for support. Inevitably, there was reference to Buckminster Fuller's geodesic dome and what appeared to be its microscopic model, a virus protein.

The sand dollar became the focus of an intense debate. In contrast to the domes that were successfully copied, the sand dollar was described as extreme, even aberrant and mysterious. The use of sea urchins as models for architecture dates back to the classic Doric columns of Greece, when the rounder urchin was used as a blueprint for the molding under the abacus of the capital. Yet the sand dollar modified its original blueprint with its slender dome. Were there advantages to this that might be copied? The slender form is not a quirk, since it has been around for millions of years. Some forms are less slender than others, with bulbous blossoms. Does it represent a possibility space that has yet to be used in human designs, or is it, like the Sydney Opera House, a dome that requires secondary support? To this crowd, secondary support meant waste, additional expense in terms of construction and material, with none of the elegance in design that denotes economy.

The first part of the debate swirled around the sand dollar's environment, diet, and behavior, which was predictable, since these influences have long been used to explain forms. But the intensity arose when the form was studied for its elements of construction. Whether its shell was a dome or a pneumatic structure engaged a biologist and a paleontologist, both of whom spoke engineering. The hydrodynamic forces of tidal shallows were considered, along with the stability of sand ripples, and the critical velocity needed to dislodge a dollar. Lift and drag figured, as did tension and compression.

Unlike their fully vaulted cousins that live in deeper waters, sand dollars hug the shore, buffeted by waves and pulled by the ebb tide, staking their claim on a different food source by burrowing. The thin dome enhances their position, as do "weight belts" of sand. Baby sand dollars take on big grains, and in addition to sand as ballast, the adults, as mentioned, use their keyholes, or lunules, as lift spoilers. The priority of ballast led to a transformation of their spines, no longer used for defense, and not so stiff or prickly as those in a fully domed urchin. They work as sieves directing the grains into food grooves, yet the food bits that might be attached are almost incidental by volume, the bulk used for ballast. Form follows function to such an extent that the dimension of the spines matches that of local grains. The sand dollar might have originally been so named because its color and texture are such perfect camouflage on a sandy beach, but the grains themselves are its currency for diet and stability. In the words of Canadian biologist Malcolm Telford, the sand dollar has perfected "the art of standing still."

Among those in the audience listening to Telford was German paleontologist Adolf Seilacher, who had compared the sand dollar to one of Otto's architectural projects for an arctic environment. The project was known as "Ice City." The first stage of construction relied on pneumatic structures, or "pneus," as Otto called them. A pneu is a membrane supported by gas, often air. A blimp is a pneu. A blowfish is partly pneu. Most fish have a swim bladder that is a pneu, for buoyancy. A soap bubble is a pneu. Otto used big balloons, inflated and sprinkled with water. When an icy crust formed, the inner pressure was released, and doors cut into the instant domes.

Seilacher found this transformation intriguing and compared it to the organic. Chicken eggs also make a transition from a pressurized membrane to a rigid structure; the shell is mineralized calcium. So is the test of the sand dollar, composed of calcium plates that fit together snugly like the shell of a tortoise. Seilacher was looking for models to explain the elements of construction and the transition of growth. It was rarely a perfectly fungible trade: a living thing grows from small dimensions, there are modifications in time within a lineage, and organic parts are frequently multifunctional, unlike most human designs. Yet his examples were piercing.

During the sand dollar's evolutionary history, its building blocks of calcium plates changed in shape from a rhombus to a hexagon. The rhombic plates allowed flexibility, like rhombic fish scales. So the primitive sand dollars of the Paleozoic were capable of collapsing, Seilacher reckons, without interior pressures akin to those of a balloon.

The new and improved plates assume a tighter fit. The sand dollar had to increase its production of calcium to build this firm rooftop. The transition led Seilacher to describe the sand dollar as a "mineralized pneu." He took it one step further, saying the tethers that held the dollar's organic "balloon" had mineralized into supportive struts for the dome. It was a wild idea, and Telford devoted many tests to disproving it (a pun to the assembled, since the surface of a sand dollar is called a "test"). A real pneu has a flexible fabric held in tension, while the supporting air is in compression. The test of a sand dollar is rather stiff, and no compression showed up on his dials. Pressures from the interior fluctuated around a zero differential and even sustained negative pressures. It is possible, Telford finished, that neither the pneu nor the dome model could explain the growth of the sand dollar and maybe it was simply a matter of growth vectors, the angles of attachments between their hexagonal plates.

The two men volleyed for the final word. Seilacher had the louder voice, but Telford had the podium. While Telford said maybe they were neither pneu nor dome, Seilacher said maybe they were both. There was no resolution. Later, when I asked Dolf Seilacher why he had given so much study to the sand dollar, he said, "Only by extremes do we learn." It was a key to this new terrain, and a new science of form.

Drawing blueprints from nature is nothing new. During the fifteenth century Leonardo da Vinci designed sleek ship hulls based on the movements of fish in water. His notebooks are rich in comparative anatomy and drawings of flying machines, many patterned on birds in flight. The

Wright brothers devised stabilizers after the way a turkey vulture employs its primary feathers to reduce turbulence at low speeds. The cockpit of the supersonic Concorde was designed to be lowered on approach, like the head of a swan. The interior braces and struts of an eagle's wing bone resemble some down-to-earth creations, particularly railroad bridges. From the fish bones for a bouillabaisse you could construct models of all of Manhattan's bridges.

Alexandre Gustave Eiffel was principally a bridge builder, and his notion of using lightweight latticework resulted in the celebrated tower he built in Paris in 1889. Eiffel honed the material to a minimum; a theoretical meltdown of the wrought iron structure, 300 meters in height, would result in a 2-inch sheet of iron at the base, covering four acres. Frei Otto also used a lattice configuration for bridges, employing the so-called fish belly design. Latticework for London's Crystal Palace was inspired by the veins and ribs of the water lily *Victoria regia*. The giant lily has leaves five to six feet in diameter, strong enough to support a 5-year-old, as the Palace engineer Paxton demonstrated with a photograph of his daughter.

While the organic world appears to express a flourish unrestrained, frugality is key. Charles Darwin wrote that natural selection is "continually trying to economize every part of the organization." Form Follows Function was a concept touted by architects (such as Lewis Sullivan and Le Corbusier) but Darwin had already discerned such a principle in his study of orchids, brilliant invitations for pollination by insects who might otherwise fly by these bee-sized landing strips, marked by directional lines and hues. Mammoth whales are better streamlined and maintained than the ironically named Very Large Crude Carriers (VLCCs), vessels that fall apart with embarrassing regularity when confronted by the same forces at sea. Optimization is seen as a natural bent, to such an extent that it led to a school of Vitalism, the view that forms were programmed to perfect themselves, as if they could anticipate the future. This predetermined and purposeful design did not acknowledge mechanical aspects such as tension or compression, the constraints of growth and materials, or that a sand dollar is slender because that form works best on the beach.

In the school of Vitalism, form was predetermined from the very beginning. Consequently, "optimal" became a loaded word, along with "design." Like the sand dollar, most organic forms have a series of ancestral fossils that document several retreats to the drawing board, and the Creator is often described as a tinkerer rather than an engineer.

That said, there's been a long time to tinker, and few human inventions show more finesse than many of the natural forms that surround us, as if they were here to impart instructions to the newcomers. Castles of clay built by African termites are better thermoregulated than any of our skyscrapers; interior chambers are maintained at a constant 78 degrees Fahrenheit. The so-called compass mounds in northern Australia are better oriented than many wanderers on their walkabouts—thus the name. The north-south alignment takes advantage of solar energy.

Trees invite particular envy. Their leaves are highly efficient solar panels, yet they accomplish more than one function. Leaves also fold to avoid drag when the winds blow, and by transpiration, move small rivers up trunks as tall as 120 meters. As the oldest living things on earth, trees also win the prize for fatigue resistance. The oldest in North America are bristlecone pines in the White Mountains near the California-Nevada border, 4,600 years old. Baobabs on the African landscape were drawing sap via a complex hydrological system before aqueducts were built in Rome.

The shell of a pearly nautilus encases what amounts to a jet propulsion engine with soft parts, less likely to break than rigid ones. The shell itself is capable of enduring depths of 800 meters, a model of maximum strength employing minimum material.

That an engineer or architect can select a material you would assume to be an advantage. Yet human inventions tend to be rigid and heavy; plant and animal materials are generally lightweight and elastic, favoring tension forces rather than compression. Things that appear to be solid and strong (concrete, bricks, glass) are often brittle. Cinderella's slipper would not serve her nearly as long as topsiders, because in terms of strength, leather is tough enough to give a little.

J.E. Gordon, professor of materials science at Reading University,

U.K., suggests that "strength for weight, metals are not too impressive when compared with plants and animals." Of course, plants and animals have to be malleable, because they grow and change. A certain malleability in human designs is desirable: buildings in earthquake zones, for example. As Roger Bacon suggested in the thirteenth century, "Nature can only be mastered by obeying its laws," a concept revived during the 1993 flood of the Mississippi, when the U.S. Corps of Engineers had to rethink their notion of rigid containment.

Life on earth is the longest trial and error experiment on record. A human engineer operates from relatively limited experience, in limited numbers, under limitations imposed by hand-wringing in the legal department. Tests can be applied to materials and small scale models, but the safety factor calculated into construction plans is also known as the "factor of ignorance," in the words of Professor Gordon, who wrote *Structures, Or Why Things Don't Fall Down*, explaining why more than a few things did. There is no way to test the extremes a structure might need to endure until that particular structure endures them. To protect their firms against the "factor of ignorance," in the past many engineers simply overbuilt, adding costs and weight.

In 1961, Frei Otto founded the Institute for the Study of Lightweight Structures, originally headquartered in Berlin. Otto, now 70, was disposed towards the lightweight from a very early age, when he made model planes out of balsa wood. At 15 he flew a glider and a few years later joined the German air force. During World War II, he was shot down and captured near Chartres; his stint as a POW was to have a major influence on his career. Otto led a construction team that repaired bridges. Supplies were limited, which forced him to work with very little material. He found that by increasing the number of tension members, concentrating compression in a few short struts, he could reduce the volume of materials he needed.

Otto eventually gained fame doing more with less in an era when less was more. Lewis Mumford described the trend as "the architecture of the future, light, aerial, open to sunlight, an architecture of voids rather than solids." Many of Otto's designs looked like tents. Skeptics took one look and made a note to check the weather forecast. Part of the problem was semantics; lightweight implied flimsy.

Weight dominates structures that rely on compression: buildings of stone, brick and concrete. Compression presses; tension pulls. It was tensile strength that appealed to Otto. Bucky Fuller gave this integrity a name: tensegrity.

The strength of a structure or material is measured by the load that will break it. A spider's web is remarkably strong, capable of not only supporting the spider, but its prey, windblown things, and the occasional mate. The cellulose of trees has extraordinary tensile strength, measured in psi (pounds of force per square inch). The tensile strength of wood, for example is 15,000 psi; the tensile strength of muscle tendon is equal to that of hemp rope (12,000 psi). The tensile strength of a spider's silk is 35,000 psi, and by weight, is five times stronger than steel. A single strand of spider's silk could stretch 50 miles before breaking under its own weight.

While no one suggested that the World Trade Towers should have been the World Trade Orbs, spider silk factories are in the works for the Monsanto company, a major U.S. producer of fibers and chemicals, and for several biotech labs in California. The genes responsible for the silk have been isolated and introduced to bacteria for a high yield. Potential products include coating on implanted heart valves, artificial tendons and ligaments, and wrappings for serious burns. Other Smart Materials are being developed to repair cracks in concrete bridges as they occur, imitating the mineralization that the sand dollar employed.

Drawing blueprints from nature is yet another argument for biodiversity, since the unknown and untested may be just as useful as the material from a spinneret. While the strength does not beat the psi's for synthetic fibers like Kevlar (used for bulletproof vests), the latter simply don't have the elasticity of silk, which can be stretched to 130 percent of its original length. The organic has an additional bonus: synthetic compounds don't biodegrade, and spider silk is naturally waterproof.

So far, this sounds like a manifesto of common sense, ghosted by Al Gore. Yet most of us are as rigid in our thinking as our buildings reflect, recycling is thick with bugaboos, and it's difficult to propose reinventing the wheel as long as gasoline is cheaper than bottled water. A logger has totally different designs on trees than Claus Mattheck's. On the other hand, Mattheck's research has been used to benefit the timber industry (for example, how to avoid splits when felling), the sonar capabilities of bats helped refine radar, dolphins have been studied by the navy for their efficient locomotion in the water, and the strength of silk is familiar to a paratrooper.

Research in this field has focused on military applications, agriculture, artificial implants, and sports medicine. The Nuclear Research Center at Karlsruhe, Germany, anticipated a transition to post–Cold War employment of physical data, and profit. This is not so much original thinking as adapting to change, which is what evolution means. The U.S. lagged behind because for twelve years, the White House saw conservation as a burr in the saddle rather than a potential model for Velcro. It was, as former President Bush would say, a Vision Thing. In September of 1993, President Clinton announced that military data on superstrong and lightweight materials would be supplied to the Big Three U.S. car makers, along with Stealth bomber and Star Wars secrets, the latter including high-speed rotors that could be used to store energy. The idea was Vice President Gore's, the goal, to develop a car with three times the energy efficiency of current models.

In this era of diminishing natural resources and gargantuan waste, nothing could be more timely. "Imagine if we could make a car that worked like the human body," anatomist Alan Walker said to me years ago. "We'd save billions in oil!" It sounds farfetched, but flight is now so trivial that it no longer occurs to us that imitating birds was once a distant dream and those who pursued it were thought ridiculous.

◆ ◆ ◆

THE 1991 Natural Structures conference in Germany was the first public symposium organized by *Sonderforschungsbereich* ("Special Research Areas"), or SFB 230, which had the ring of *Fahrenheit 451*. The 1967 film by François Truffaut, based on the Ray Bradbury book, has a final scene that is apt. The erudite Julie Christie takes to the woods in a revolt against book burning. She is joined by a converted fireman (played by Oscar Werner) whose lawful task was not to extinguish fires but to start them, torching literature (451 degrees Fahrenheit is the temperature at which paper burns). In a communal retreat of book people, endangered classics are committed to memory and the words of great thinkers shared over the campfire. There is a moment, in the snowfall, when you see only the trees in the forest yet hear the words of the book people. As the fireman is being introduced around, people step forward, saying, "Hello, I'm Charles Dickens' *A Tale of Two Cities*," or, another, "Hello, I'm *Alice in Wonderland*." The host: "Yes . . . *Through the Looking Glass* must be around here someplace."

The University of Stuttgart conference, held on the wooded campus in Vaihingen in October, did seem to be a cloister of intelligentsia, citing from books and scholarly papers. It was largely European (of several hundred participants, only four attended from the U.S.; I was the only American reporter). Prior to this, there had been satellite projects, such as SFB 53, on Construction Morphology. SFB 83 focused on the Lightweight, SFB 64 worked with Nets. Project numbers related to chronology, and the bulk of funding for the gatherings came from the German government and local corporate foundations. Sessions were held simultaneously in three separate lecture halls, and exchanges continued over dinner and well into the schnapps, to resume at breakfast—easy since everyone was clustered in a small group of hotels. The energy was exceptional and the mood that of a Think Again Tank.

The Think Again Tank has momentum, which may be contagious and carried by airborne spores. The innovations of Claus Mattheck were initiated and pursued independently of the work begun by Frei Otto. SFB has a sister group in England, known as SEB (Society of Experimental Biology). SEB focuses on biomechanics: how muscles and bones work, the energy of cracks and composite materials that halt them; the locomotion of horses and cheetahs and humans; the dynamics of flight and the evolution of the wing; and engineering principles of plants.

Until recently, biomechanics enjoyed less government support in the

U.S., although a little hotbed of research at Duke University became influential, as was the intention. The introduction to *Mechanical Design in Organisms* stated, "This book is frankly evangelical; we wish to modify biologists view of the world."

The vision they expand belonged to a man whose name was mentioned with amazing regularity at both the German and British conferences.

D'Arcy Thompson (1860–1948) believed that math could explain the mysteries of life. The Scottish biologist looked at nautilus shells and kudu horns and compared their logarithmic spirals. The hexagonal patterns of honeycombs and gossamer wings intrigued him, as did the bulky bones of an elephant, contrasted to the latticework of bones in birds and fishes, less burdened by the forces of gravity.

He drew geometrical lines on organic forms to connect their central features, much the same way Gothic cathedrals, for all their ornateness, are traced with triangles and circles, or the growth of the human body is demarked at the navel by the so-called Golden Mean, a proportion said to be so pleasing that it determined many human designs, from pyramids to playing cards.

Thompson's 1917 book, *On Growth and Form*, suggested that the shapes of all living things are largely the result of physical forces—gravity, viscosity, tension, and compression. These were engineering terms, the stuff of physical laws, distilled into numbers. Math was the tool of chemistry and physics. Biologists preferred anatomical terms and granted Latin names to species. Forms were seen as the result of natural selection. Shapes were related to behavior and diet, not physics.

At the time, Thompson's book struck many readers as a radical way of looking at natural forms, although he had simply drawn together some ancient facts and pondered them in a novel way. He wrote beautifully about everyday matters, from the surface tension that keeps a drop of water dangling from a faucet to rain drops that marry in a free fall. "They rotate as they fall, and if two rotate in contrary directions they draw together. . . . This only happens when two drops are falling side by side, and since the rate of fall depends on size it always is a pair of coequal drops which so meet."

One of the best chapters in Thompson's book is "On Magnitude," which considers dimension and scale (a subject he deals with "magnificently," punned G. Evelyn Hutchinson of Yale). If a mouse were to grow to the size of a rhino, would its legs (larger by the same proportion) support it? Thompson used a basic formula for growth established by the Greek mathematician, Archimedes: Surface area varies with length squared, while volume varies with length cubed. Dimension and scale not only reflect measurable aspects of forms—why an elephant's legs are so thick, and those of a giraffe thin—but also influence behavior and longevity. Small animals lose heat rapidly because of their relatively large surface area, and this partly dictates their need for greater consumption of calories. A hummingbird lives a short life of great intensity. Larger animals tend to consume less food, less oxygen, and move about more slowly; their pulse is slower, they live longer, and their volume generates greater heat in proportion to their surface area. Being large also has its price. Were a bird to double in size, its energy requirements for flight would increase substantially. Large birds save energy by soaring, an option unavailable to the hummingbird. All these forces configure.

Evolutionary biologist Peter Medawar describes Thompson's book as "beyond comparison the finest work of literature in all the annals of science that have been recorded in the English tongue." *The Whole Earth Catalog* listed it as a "paradigm classic." Yet recognition was belated. In his book *Chaos*, James Gleick wrote that "D'Arcy Thompson surely stands as the most influential biologist ever left on the fringes of legitimate science." Harvard professor Stephen Jay Gould reckons Thompson's ideas "have gained new impact in a science that only now has the technology to deal with his insights."

The new technology includes the scanning electron microscope (SEM), which reveals intricate patterns of organic materials; advances in chemistry and molecular biology help explain the building blocks. The most influential tool is the computer, which allows rapid calculations and playing with the parameters of three-dimensional forms. Prior to this, some of Thompson's ideas were considered "analytically unwieldy."

Now logarithms advance with the lubricity of a laser printer, replacing the slide rule's handy organization of powers of ten.

The slide rule, or wooden slipstick, was a symbol of engineering, although it was equally employed by mathematicians and physicists. By sliding a smaller ruler within a larger one, roots and quotients could be aligned, top to bottom, and logs figured by addition and subtraction. There were also round graduated tables that spun. The slide rule, first manufactured in the U.S. in 1891 by Keuffel & Esser, now gathers dust alongside the abacus and the sextant. It is past tense. A young German physicist, when told the title of my book, asked, "What's a slide rule?"

Some aficionados who lament its passing are reluctant to trust the dizzying calculations of the "idiot-savant computer." Henry Petroski, of Duke, who preferred K & E's Log Log Duplex Decitrig as his "silent partner and constant companion," has written that the "limitations of the slide rule were also its strengths." He argues that the absence of decimal points on the slide rule forced the user to maintain a cerebral grasp of magnitude. Of course, this is also an engraved invitation for human error, but he thinks computer calculations do not demand the same "reflection." Petroski, who favors traditional tools to such an extent that he published a 454-page book on the pencil, has a point.

"Finite element analysis and even computer simulations can tell us nothing about the growth mechanism of how sea urchins develop into the shapes which we see," warned Malcolm Telford. "Unless we recognize this limitation, biologically inclined engineers and mechanically inclined biologists risk the fate of our great forerunner, D'Arcy Thompson," left on the fringes.

Do we use numbers to organize our world, or do they organize us? Science thrives on discerning patterns and threading them together into a neat story. So did the legend of the sand dollar. The planets in our solar system seemed hopelessly disorganized until Isaac Newton came up with the sweeping simplicity of a formula for gravity. It worked, but often an elegant idea is favored because of its neatness. For years

Johannes Kepler ignored the data that put the planets on an elliptical path about the sun, because the neatness of a circle was considered sacred. The human mind tends to organize complex things into artificial and abstract categories as a way of dealing with reality, and rampant innumeracy suggests that we reflect very little on the numbers that surround us. Numbers, like words, are a tool, yet for all the pliability of language, words occasionally fail to convey the complexities that nature presents to us, as when light was discovered to be both particle and wave.

D'Arcy Thompson reflected on ancient approaches to math and examined the origins of numbers as tools. He applied the precision of numbers, and delved the geometrical aspect of forms, playing with models. Yet models lack any warts, and there is some truth to the dichotomy: The devil is in the numbers, and God is in the details. Consequently, Claus Mattheck made his entrance with a tree limb that had failed. It was an example of a blueprint not to copy.

A famous pilot who earned his wings in literature, Antoine de Saint Exupéry, wrote of the evolution of airplane design in *Wind, Sand and Stars:* "Have you thought, not only about the airplane but about whatever man builds, that all man's industrial efforts, all his computations and calculations . . . invariably culminate in the production of a thing whose soul and guiding principle is the ultimate principle of simplicity? It is as if there were a natural law which ordained that to achieve this end. . . . In anything at all, perfection is finally attained not when there is no longer anything to add, but when there is no longer anything to take away."

Claus Mattheck put it differently: "The idea," he said, "is to get rid of waste."

2. The Hazard Beam

*Since the weight of a fruit increases as the cube of its linear
dimensions, while the strength of the stalk increases as the square,
it follows that the stalk must grow out of apparent due proportion
to the fruit: or, alternately, that tall trees should not bear large
fruit on slender branches, and that melons and pumpkins must lie
on the ground.*

— D'ARCY THOMPSON

PHYSICIST Claus Mattheck became interested in the "body language" of trees when he encountered one more or less in the shape
of a question mark. The tree was on a beachfront on the coast of France,
about 20 miles south of Bordeaux, where it had survived fierce storms
off the Atlantic. Its trunk had a zig and zag of unequal length and dimension. It looked like a map of a river deformed by dams.

While most investigations rely on tree rings, Mattheck pondered the
je ne sais quoi of this pine's exterior, so unlike the other forms he surveyed.
In German forests, as in most dense stands, trunks tend to rise lean and
tall in general unison, somewhat protected from the wind. This French
pine was alone and something of a natural bonsai. In addition to its contorted trunk, many branches were kaput. That it was ugly is beyond
question, but trees have a way of growing out of it.

A tree is shaped by the loads it must carry. "The tree holding out its arm . . . is a great matter," wrote Buckminster Fuller; "If you try lifting a fifty pound suitcase while holding out your arm horizontally, you'll find you can't do it. But a tree is often holding out a branch weighing as much as five tons. Holding five tons out there horizontally, and being able to do so in a hurricane, is perfectly extraordinary."

The massive crown of an oak requires a thick trunk to support it, while a pine (the quintessential conehead) can grow lean and tall. Contours of the trunk are also shaped by their burden, which configures the depth and spread of their roots more than water; the length of roots often equal the tree in height. Local conditions such as prevailing winds can alter the shape of a tree, as can herds of giraffe, but such alterations are ephemeral. The basic blueprint is informed by gravity.

Like the formulas for gravity that Newton devised (said to have been inspired by an apple that fell in his mother's orchard) there are formulas for the growth and form of parts of trees. Many are ratios of the cubed to the squared, including theoretical limits to the height of any tree. Others define the dimension and density of a Granny Smith apple, designed to fall to earth at the ripe moment. Ideally, this is the same moment its stalk simply fails to support it. A good wind can alter the delivery schedule. A bad wind can alter everything. Extreme loads of ice can create telephone poles, but burdened by snow, trees can spring back into shape, and have a remarkable ability to repair themselves, to replace broken limbs, even to right themselves in an affront to gravity.

Mattheck plumbs the load history of trees, their maintenance program, and their preventive health plan. They can not anticipate, but their form has incorporated the lessons of history. We might react faster, but trees have been reacting longer. Their health plan includes healing bark wounds to prevent the entrance of bacteria and fungi that can cause fatal diseases. Maintaining a balanced load is essential to avoid a stressed out, dysfunctional core, known as brittle heart.

A healthy tree is centered and said to be geotropic, responding to the gravitational forces at the core of the earth. This is why a sapling on a mountain has an inherent inclination to grow away from the center of the earth, rather than perpendicular to the slope. Trees, like humans, work better when maintaining good posture and an alignment of parts. Their hydraulics can lift water 300 feet on a daily basis, inspired by leaves that transpire in their tiny pores, pumping until the sun goes down. Flow is everything, including the strains of everyday life. This is what Mattheck was keen to measure, and in this sense, the pine was perfect.

Mattheck considers D'Arcy Thompson a pioneer. The eighteenth-century biologist called his book *On Growth and Form* because he considered the two factors "inseparably associated," and the link provides a key for a new way of seeing things. For example, when a storm throws a trunk off balance, the tree works to get its center of gravity directly over the rootstock again. If the leading stem is damaged, the tree works to pull in a side branch as a replacement. The only way a tree can move to the center is to grow in that direction.

To accomplish this, trees generate what's known as reaction wood. Softwoods (like pine) grow new wood on the outside, compressing the trunk towards the center of gravity. Hardwoods (like oak) favor growth on the inside, creating a tension that pulls the trunk back over the center. This reaction wood occurs in the cambium, a ring of cellular tissue just inside the bark from which new wood is made, and it kicks in when sensing strain in a particular region. Hormones play a role, but calcium also figures. The reaction is fairly prompt, but the growth takes time, several seasons of spring and summer regeneration, easily moving beyond a decade. So when Mattheck studies a living tree, he has a mere snapshot of a work in progress, and in the case of the French pine, a gnarled history.

Mattheck noticed a couple of oblong notches on the pine. They mark the site where a branch failed. These wounds look like eyelids of darkness, elephant acne scars. The older the scar, the more difficult to discern its cause. Frost, for example, can create a thin crack, or rib, that runs for several feet down the bark; freezes can create cankers. Most notches are oval depressions with an elliptical trim; the healing lips rise like the threaded edge of a button hole and work to close the hole with

A

B

In image A, the phototropic overcame the geotropic needs when a tree grew at a right angle to gather sunlight in a sanctuary in Central Park known as Strawberry Fields. In image B, a limb in Manhattan's Riverside Park succumbs to gravity but continues

C

D

to live. *In image C, the remains of a Brittle Heart in Strawberry Fields, a result of poor distribution of stress that concentrated in the core. Image D, a* Euphorbia robecchii *in Kenya, with limbs demonstrating the curved growth known as epinasty.*

enveloping ridges. The original stem of the French pine was damaged; it fell, and was replaced by a new branch. That replacement branch, under the influence of reaction wood, grew towards the center. Then it was broken, and a third branch from the other side was recruited, and it grew toward the center. The result was a meandering trunk, with detours marked by notches.

Failed center stems can also be replaced by two branches, which results in a fork. Reaction wood can incline a shaded tree towards sunlight, creating a stringy set of branches on one side, a configuration known as a harp. Reaction wood does not react to obstacles so much as it envelops them.

Mattheck found trunks embracing boulders, road signs, fences, garden statues, even a crucifix. He found a conifer that had overtaken its neighbor, parallel trunks that had merged, and a beech that perservered in a horizontal position (horizontal harp). Branches that grew back into the trunk he called jug handles. Friends learned of his interest and sent him photographs of kinky trees; he was led down many paths in England, in British Columbia, in the U.S. in Idaho and Maine, and along the edges of deserts in Namibia. After a severe storm devastated an ancient forest on Queen Charlotte islands north of Vancouver, Mattheck braved local bears and the stench and slimy patina of the end of the salmon migration. His mission was to measure fallen trunks.

That he traveled such a distance and endured such conditions ("We cannot bathe as often as we'd like, because the waters are full of dead salmon") suggests an inordinate interest in failures. But only from extremes do they learn; to know the perimeters of a tree's safety factor, Mattheck needed failures. Since every tree is so different, and the conditions that affect even the same species are wildly variable, he needed lots of failures.

He seeks the crooked and uprooted, the bent out of shape. Cracks intrigue him, and he often returns to the same sites in the German Palatinate to study their progression. To Mattheck, "the tree which had to fight for its survival will always have the more inspiring appearance." He compares his observations to that of "judging character from the features of a human face," where lines of stress lend a furrowed brow or crow's feet, making you wonder if a downturned smile is the result of a brittle heart or some unbearable gravity. That these images are only metaphorical he makes clear, emphasizing the precision of biomechanics. The forces of stress on a tree can be measured with exactitude, the interplay of tension lines that run up and down the sides of a trunk compared to the compression forces of the core, the turgor of cells, or the elasticity of wood fibers (even if the tree takes a siesta after lunch—trees can wilt in the heat of the day to conserve water). He is scientific, tapping the trunk with a hammer that senses resonant impulses of the interior, boring with a digital "fractometer" he invented, but even an amateur can hear the hollow tap of an empty core. He does not ignore tree rings or any angle of data. He photographs trees, and computes their variations into like parts—treatise on trunks, leitmotif on leaves, notations on notches. Yet inescapably Mattheck has come to see trees as individuals, and in addition to his vintage idiom of "body language," he describes his drawings in an anthropomorphic way, one of them captioned "A Young Beech Braving Adversity."

Mattheck's playful black and white renderings feature in *Trees, The Mechanical Design*, published in 1991, a work followed in 1993 by *Designs in Nature*, which includes his "philosophy on the relationship between mankind, tree growth and energy." He adds, "For this I created the expression, ecological design." Eco-Design popped up in many places at once because it worked like a cliche, with the impact he intended. He appears to feel privileged to do what he does for a living, and ends his most emphatic statements with a chuckle. His handwritten faxes are peppered with exclamation marks, and at 47, he belongs to the generation that refuses to look middle-aged. In addition to his wire rims and leather attire, Mattheck has a John Lennon nose. Were you to pass him on the street, you might assume he was in the rock and roll trade rather than a licensed tree hugger. He is not a dendrologist, merely studying trees, but a physicist in deep forest. If his analogies are somewhat romantic, he is shrewd as a salesman, which may have something to do with once having had a price on his head.

Mattheck was born in 1947 in Dresden, where he received a Ph.D. in theoretical physics. At the age of 28 he attempted to "escape in a rubber boat at sea," but was captured by East German officials, who kept him in prison for two years.

Then, in a stroke of supply-side economics, Mattheck was "sold" to West Germany by his captives in the East and released from prison, perhaps the only human being delighted to be selected for auction. Mattheck was hardly enslaved in his new home but gained the bonus of a salary with his freedom. With his wife, Marita, and son, Willie, he maintains a home west of the Rhine, on the edge of the Palatinate, and "far from the cities!" His stint at the Karlsruhe Nuclear Research Center began in the department of Fracture Mechanics.

The study of cracks considers impact and fatigue in the bones of horses and ballerinas, the fuselage of aircraft, windshields, the transport of eggs, the struts of artificial heart valves, and so on. There are formulas to define a critical crack length, and numerical values attached to a material's ability to fracture, alternately defined as "creating a new surface," and "the communication of energy between the elements of a material." Fractures can disengage molecular bonds or splinter fibers; cotton fabric is easy to tear once you've made a notch. Fruits, like muscle, rarely crack, but distribute the forces in a bruise. Cracks love the brittle, yet even rubber can qualify when exposed to freezing conditions, as physicist Richard Feynman demonstrated before a U.S. Congressional hearing, simply by dropping an O-ring of the type used in the ill-fated space shuttle Challenger into a glass of ice water.

Wood is a composite structure, combining the features of two or more materials, common in organic forms, and wisely copied. Like fiberglass, wood has reinforcing fibers in a matrix that by itself would prove brittle. An even more advanced combination is graphite fibers in epoxy, used in the Voyager airplane that glided around the world in 1986 without refueling, and in 1993, on McDonnell-Douglas's DC-X prototype for a cost efficient space shuttle, canned by U.S. Defense Department officials because the "Buck Rogers technology" sounded too good to be true.

The strength of fiberglass is such that in its early use, its safety factor was undercalculated. Twice the material needed was used (for example, in boats built in the seventies), before experience proved that half the thickness would suffice. The glass fibers absorb pressures, and can arrest a crack by their configuration: helical and cross-links, circular or axial. Unfinished fiberglass looks like a sheet of white mohair. The fibers can assume some isolation within the matrix, and this remoteness does not allow for the communication of energy required for a crack. Fiberglass is said to become stronger at lower temperatures. A composite is a tough, yet lightweight, material. The same is true for wood, remarkably low in density. The greater bulk of a living tree is water. A mature Douglas fir may contain 4,000 liters of water.

Dried wood is brittle, while fresh-cut timber bends. The wood's elasticity can be extended by boiling and steaming, an hour for every inch of thickness. While the majority of cells in a tree depend on turgor, in reaction wood, cells are arranged differently, to induce forces rather than absorb them. For example, reaction cells of tension are flat; they dehydrate, crystallize, and shrink. By shrinking, they pull to correct a tree's posture. Boards cut from tension wood contract along the grain. Reaction wood is not ideal for human designs, because its design was so specific to another need. Even the knots of pine, formed by branches growing from the trunk, are vulnerable if the wood becomes dehydrated. The knots attract stress, like any hole in anything that must bear weight—a cavity in a molar, a screw hole on a mast. Stress trajectories drawn up by physicists resemble a pinstripe suit, with lines equidistant. These lines are forced to travel around notches, knots, and in doing so push distant lines of strain together, like a bottleneck, when ideally they should flow evenly. The stress around a tree wound is often three times as much as the actual applied pressure.

Notches can be useful. A little notch in the paper wrapping of sugar cubes takes the stress off the cube, as do grooved notches of pure chocolate bars. (Bars with nuts are a composite structure.) Notches appear as perforations on postage stamps, and the test of a sand dollar has round perforations positioned to halt a crack. Steve Vogel of Duke compares these holes to portholes in ships and the shapes of holes in bones: "These, to the best of our knowledge today, are the best designs to minimize crack propagation, and that's engineering."

While notch is a technical term in engineering for a focus of high stress, it is in practice often synonymous with a crack, which is, after all, a small notch in the beginning. When beavers fell a tree, they carve a big notch into the side. Beavers have discovered the weak part of trees and violate the compression of the interior. It takes an elephant or a car to push over the tensile structure. The compression strength of wood is only about a third of its tensile strength.

The cellulose of plants is described by Professor J.E. Gordon as a "tensile material without peer." The strength derives from an ability to store energy like a spring. Cellulose is a polysaccharide of high molecular weight. (Synthetic materials such as nylon and Teflon have similar properties, with molecular chains that crystallize and extend into a triple helix.) In its atomic structure, cellulose is like other components in organic forms with tensile strength—such as collagen, which provides elasticity for our skin and forms the links between the hexagonal plates on the sand dollar test. Wood fibers work like a rope or a cable. When pulled, the fibers flex elastically and recover. The smooth flow of tensile strength is interrupted when a rope has a knot in it, just as the elasticity of a muscle is crippled by a spasm, a knot where oxygen and blood aren't flowing.

Cellulose dominates the cambium just beneath the bark, a circle of cellular flexibility that allows a tree to respond to the wind, and recover, creating a condition known as prestressed. When seen in a cross section, nearly ninety percent of these cells are arranged in a honeycomb structure, the hexagonal packaging that occurs repeatedly in nature and results in lightweight strength. The weaker parallel arrangement along the grain is why wood is easier to split than to chop across. That said, the ability of living wood to halt a crack is, weight for weight, better than most steels. So a failure is always interesting, and the splintered limb that Mattheck employed as a visual aid in Stuttgart was such a failure. He described it as a hazard beam.

The word "beam" derives from "tree" in Old English, and in Sanskrit, "root," or "to grow." Ceiling rafters were often composed of whole trees and still are in the rustic. Beams replaced arches by spanning flat across, transmitting the weight of the roof down into the walls with comparative finesse; walls could be thinner with no need for buttressing. A beam supported only on one end, such as a tree limb, is a cantilever beam.

Cantilevers were the signature of Frank Lloyd Wright, inspired by trees to such an extent that he instructed one of his students to stop reading textbooks and draw trees; one "who could sketch the forms of trees, with their characteristics faithfully portrayed, will be a good architect." Wright devised cantilevered lamps, extending from the wall of a room, and cantilevered verandas for the famous house known as Fallingwater at Bear Run, Pennsylvania. According to a biography by Donald Hoffman, "Wright believed the cantilever to be the most romantic of all possibilities in structure, and he made the cantilever his main instrument for asserting a new freedom in architecture."

This freedom made the construction crew for Fallingwater somewhat nervous, and when Wright instructed them to remove the frames for the bold concrete verandas, there was hesitation. When the frames were removed, the cantilevered extensions settled slightly, and they remain aloft to this day, a stunning sight, since a beam can sag even with support at both ends. This deflection occurs in proportion to the cube of the beam's length. An empty bookshelf 8 feet long will sag 16 times as much as a shelf 2 feet long. For bookshelves or bridges or rustic lodges, support struts take the thrusts of deflection. In lieu of such struts, an obvious reinforcement is to be thick at the base and taper, which is the form assumed by limbs and a tree trunk. The loads of a tree travel to the base.

Wind loading causes maximum bending moments at the thickest part of the tree, the butt. "On the other hand," Mattheck notes, "it causes a zero moment at the very top of the tree where thickness approaches zero." Approaching zero is more likely in winter, when deciduous trees shed their leaves. In trees with full crowns, leaves fold to avoid drag. Broad leaves tend to curl up into a streamlined cone or cylinder. Steve Vogel tested various leaves in a wind tunnel, and his photographs of the leaves of tulip trees taken in 20 to 40 mile per hour winds look like instructions for rolling a cigar.

A common reinforcement of beams in the organic is to bundle, which occurs with celery. To some extent the same is true of bamboo, which grows in clusters, and supplies excellent fishing poles. (Bamboo also has reinforcing ribs, nodes that resemble bulkheads in ships and airplanes.) On a smaller, microscopic, level, wood fibers are bundled and concentrated at the connective base of a limb, a branch, a leaf stem, or a fruit stalk. In a coconut palm, you can see these fibers at the base of a frond, especially around old ones on the lower part of the trunk. They loop around the trunk itself, then extend at an angle and spread with the look of woven fabric. The coconut palm grows exceptionally tall and survives exceptional winds although it scarcely confronts hurricanes with a 5-ton arm. The frond is designed like a feather, with thin, pinnate leaves that diminish drag, and the weight is usually less than 10 kilograms.

With most tree limbs, the upper side grows faster, which lends a curve. This same subtle curve occurs in dandelions, onions, and wheat and some other grasses, with hollow tubes that would buckle but for the cellulose along the wall that provides support. The tension cells shrink to allow recovery when the wind blows, or in grass after it's been trampled. The upper half of a tree limb is pulled in tension, the lower, pressed in compression. They work like our muscles do with bones, a compression item, and brittle. In most organic things, there is this relationship of tension and compression—even with the spider's web, the spider and its prey providing compression.

Muscles, of course, stretch and contract, and their ability to recover from carrying a load reflects their strength. Exercise creates a condition in which muscle fibers are prestressed, which is what happens with the cellulose fibers of trees and all plants. It's as if the fibers had a memory, and in a way they do: been there, done that. Just as muscles grow under repetitive strain and reflect the loads they carry, so do trees.

Most plants subjected to wind grow stronger stems, and even trees inside a greenhouse respond with a thicker trunk if given a mechanical tug. Tomato seedlings grow stronger stems if shaken a minute a day. Trees supported by guy wires can collapse when the wires are removed, since they did not develop the vitality of response that local wind conditions normally cultivate.

The trunk of a spruce or pine can be prestressed to such an extent that it maintains a postmortem elasticity, serving as a spar for sailboats, where it continues to deal with loads of wind. The prestressed condition of trees was so attractive to boatwrights that they sent scouts into the woods to find sections of trees that naturally qualified for the cheek knee, the stern post, the quarter deck hanging knee, the crutch, the web frame knee, the futtock. In the mangrove forests around the Lamu archipelago, where Arab dhows are built, the prize is known as jungle crook.

A tree blowing in the wind has complex movements, especially when crowns are broad. The branches of an oak tree are frequently longer than the tree is tall. A *Quercus virginia* near Lewisburg, Louisiana, measures 55 feet in height but has a spread of 132 feet. "It is fascinating to sight along one of these branches during a wind storm," notes Steve Wainwright of Duke, who observed an oak in action: "The branch's shape is a gentle helix and as the wind buffets it up and down and from side to side, the spring in the helix 'works' by visually shortening and lengthening. This could be an effective way to distribute bending stresses in the branch . . . that would avoid breakage."

In Mattheck's hazard beam, the curved fibers were straightened by their external load. As they unfurled and stretched, the interface to halt cracks was deconstructed. Elasticity was lost, and the wood became brittle. It survived for a long time because the upper part of the limb, above the crack, worked like a tendon for an arm bone. Then it split and fell, a sacrifice which Mattheck reckons was in the best interest of the rest of the tree, taking away an area of stress beyond repair. "It is cheaper for nature to accept the occasional failure," he wrote, "instead of making them all so failsafe by using much more material" for a single "worst case loading."

The hazard beam delivered some guidelines for drawing blueprints from nature. If trees are design teachers, they can also demonstrate what not to do. Such a hazard can be noted and diminished by human correction, with the option of studying the jillions of cantilever beams that remain successfully aloft. Too, form may follow function, but it doesn't necessarily mean a form will function elsewhere, even though Mattheck has made successful transpositions, as have others; the base of the Eiffel Tower took inspiration from the trunk of an oak, which D'Arcy Thompson also credits for the shape of Smeaton light houses. Either Smeaton and Eiffel were looking at different oaks, or their needs were different—which they were, involving considerations of height, wind conditions, building material, and function.

The minimum curvature of support at such a base has subsequently been refined and is known as a Baud curve after the Swiss engineer who devised it in 1934 to diminish stresses in a T beam. Baud curves exist in nature, for example, in the curve at the base of kudu horns, in the trabeculae of bone in a human joint, and in the two-pronged tip of a camel thorn, all of which Mattheck studied in order to refine human designs. As there is a Baud curve, someday there may be Mattheck ridges.

Mattheck found that the ridges growing around a tree wound reduce stress, and he refined a medical implant, a long bone plate, by including parallel ridges. (In keeping with his theme of getting rid of waste, rather than adding to the existing structure he cut away a rectangular channel to create the ridges on the sides.) But rather than form, what Mattheck prefers to copy is the "mechanism of growth."

Mattheck explains that trees "always try to grow into a state with constant stress on their surface." Constant stress sounds like an unhappy situation, but he means homogeneous distribution. Stress from loads is inevitable for anything terrestrial, and the trick is not to let it accumulate in one area. As far as copying the mechanism of growth, Mattheck found that stress around a wound actually dropped from triple strength back to normal after reaction wood healed over the area and created ridges. He also discovered that a tree responds fastest to the areas of greatest stress. This is what a designer wants to achieve with a blueprint.

While some compare the optimum form with perfection, there are constraints and the less-than-perfect response. For example, surface wounds of carved initials ("I Love Lucy") can create greater threats to a tree than powerful storms, probably because trees have been dealing with loads of winds much longer than with graffiti. Mattheck defines optimization as "minimum weight with a homogeneous distribution of surface stresses." This also applies to a bone, the wing of a dragonfly, or a honeycomb. He describes the constant stress axiom as "one of the most general design rules in biology," and thinks industrial applications, including automobile wheels, should strive for the same.

Tumbleweeds are one of the rare organic forms that roll along. A couple of bacteria use wheels for locomotion, but the restraints of viscosity limit living wheels above the microscopic, because it would take so much energy to move, a task often compared to a human trying to swim in molasses. Even the form of a shark reduced to the size of sperm would have a struggle with a tail of considerable length. Some improvements to reduce the drag on the automobile wheel have been achieved with General Motors' electric car, the Impact, designed by a team led by Paul MacCready (see next chapter). Fittingly, they include lightweight, aluminum spokes, and some small bacteria have radial spokes to support their tiny cilia or flagella, which substitute for wheels.

◆ ◆ ◆

With his colleague Andreas Baumgartner, Mattheck devised the Soft Kill Option, a computer program to optimize designs. It eliminates "soft," unnecessary parts of a design, and "hardens," or reinforces, areas vulnerable to fracture.

The Soft Kill Option program has been used to refine automobile parts, and clients include electrical engineering firms and a company that produces electric shavers. Eighteen licenses for the program were sold within a year of its introduction. Mattheck's staff at Karlsruhe serve as consultants, and part of the contract is that Mattheck and his team can't divulge the highly competitive results. The basics, however, are clear, and he describes several projects.

One was uniquely personal. Mattheck's leg was broken in a car wreck, and the bones were splintered to such an extent they required reinforcement, held in place by stainless steel screws. "These screws often broke under stress," he said, meaning the normal pressures upon a human leg. Tension and compression, torsion and shear. (Pull, push, twist and slide.) He needed a screw that resisted fatigue, and, like a tree, engaged his energies into self-repair.

The first step was a detailed outline of the screw expanded to a large scale blueprint. Using the finite element method (FEM), a mesh divides the blueprint into a three-dimensional grid. To simulate organic growth, Mattheck recycled a program of thermal expansion, adding area in tiny increments. Of course, tree growth has nothing to do with thermal expansion; this happened to be a feature of a commercially available program known as ABAQUS that works graphically. Then, the user incorporates formulas for elasticity (Young's modulus) and stress distribution (von Mises) based on the material. Areas of stress are automatically graded by colors.

Visually, it's as simple as a Visine coloring book; the object of the game is to make the red disappear. The user can strengthen areas of high stress by adding a layer of "growth," the way a tree adds reaction wood. These are subtle additions; the magnification of the blueprint allows changes in fractions of millimeters. In areas of very low stress, excess material can be trimmed. The curvature or angle of edits and additions is critical, so Baud curves and the like are used as guidelines. The program can be run several times with different variations until the red disappears. Then you go after the yellow, the next grade of stress. An ideal blueprint is indigo.

Playing with blueprints goes back to a 1904 theory that considered all possible frameworks within a space, and (ironically) eliminated those that might allow any deflection. This resulted in great stiffness, part of the reason the screws kept breaking. And Mattheck favored getting rid of waste rather than just partitioning a solid.

Computer optimization programs have been around since 1975. Mattheck's may be the first to employ the trade secrets of trees, but distributing stress may not solve problems of a gargantuan nature. Super tankers and giant bulk carriers are the largest vessels constructed; the largest, 375 yards long, is longer and wider than the *Queen Elizabeth II*, and heavier by five and a half times. Programs for optimizing carriers were applied to some designs and lightened the weight by several thousand tons. But an even distribution of stresses, according to Capt. W.S. Morrison of the International Maritime Organization, "meant that if something went wrong somewhere it went wrong everywhere." Often loads are complicated by poor maintenance and uneven distribution of cargo, creating a sag at midship or an arch known as a hogback. It may be that the largest carriers have breached the limits of size. The Empire State Building is 1,048 feet tall. A carrier 375 yards long would extend for 1,125 feet.

Mattheck began with small but telling forms, such as a link in a chain. His refined screw was nearly that simple. One change involved a tiny addition to the base of the threads, the same point of vulnerability as in his hazard beam. He reinforced the threads only where strength was needed, and trimmed the wasteful rest. The result was stronger and lighter, and while it didn't exactly bend and spring back like a sapling in the wind, he said it was more "pliable," which I asked him to repeat, then spell it, convinced it was not the word he meant. It was.

"After we applied the CAO simulation and optimized their design, they are now one hundred and five times stronger in laboratory tests than the unoptimized design. They are fatigue resistant." Mattheck also described the result as "more vital," as if the screw had developed a life of its own. In his view it has: "When a screw grows like a tree, it really grows."

◆ ◆ ◆

A CALIFORNIA sequoia known as General Sherman is ranked as the biggest of the big in the National Register of Big Trees, published every two years by the American Forestry Association. This *Sequoiadendron giganteum* is 275 feet tall. Its estimated age is 2,500 years. General Sherman grows at an annual rate of forty cubic feet, adding about a ton of wood each year. Most of this growth is in girth and has been for 2,000 years or so, since sequoias attain much of their height during their first 500 years. Yet the trunk never rises. Had a Clovis arrowhead been inextricably positioned at eye level 2,450 years ago, it would still be at eye level.

All trees grow from the very ends—the tips of stems and branches, the tips of roots. Trees favor growth at the top of the crown after a certain age, and to concentrate nutrition in the northern frontier, lower leaves succumb to the shade, then lower limbs are shed, no longer useful. A cantilever beam on General Sherman fell during a storm in

1978. It was 140 feet long, and 6 feet in diameter at its base, and weighed 5-tons.

To engineers, giant sequoias are the world's largest freestanding columns, monuments to fatigue resistance, and structures with a built-in safety factor 4 times the necessary. Safety factor for whom? you might ask when walking beneath one. In calculating safety factor, the material strength is related to the excess load the structure will bear; most mature trees can handle an increase of about 4.3 times above normal before failure will occur. (Small mammal bones can accommodate between 3 and 4 times the normal load, with elephant and buffalo weighing in with a safety factor of 4 and above.)

As elaborate harvest machinery, tree roots also serve as anchorage, growing thicker on the leeward side. Mature yew trees develop multiple trunks, blending the new with the old. Buttress is a misnomer, Mattheck says, as this development represents not reinforcements of compression but tensile structures "designed to act like ropes stabilizing a tent pole." The aerial roots of a fig, if pot bound, can become truly adventitous, raising the container off the floor. The aptly named *Ficus elastica*, an Indian rubber plant used like parsley to decorate dreary offices, can develop gargantuan banyans in the wild. It is a statement of flourish that masks trivial furniture design in print ads, when a well-placed palm or ficus takes the focus off the ugly sofa. Green was the first color that primate eyes discerned in the forest, and, as Emerson said, the mind loves its old home.

Trees attained their spectacular height in an apparent race for fuel. Extra fuel is recycled, with other plants consuming the nutrients of the dead in tropical forests. In temperate forests, recycling is by landfill, where forest duff becomes buried. This led to fossil fuels like coal—condensed solar energy stored away 275 million years ago in an era of such productivity it is known as the Carboniferous. The red embers of coal are inspired by tiny remains of ancient sunlight. Petroleum is still being produced in ocean basins today, in the Gulfs of California and Mexico, the Persian Gulf, the Orinoco Delta in Venezuela, and the Caspian Sea.

During the Carboniferous, other plants attempted to stay in the race to capture the sun, including horsetails and ground pine, but most trees simply left them in the shade. The oldest in the U.S., bristlecone pines, are not tall and are scarcely linear. These are not quintessential cone-heads but beauties in decline. Wizened forms survive in the mountains of California and Nevada. In the Toiyabe National Forest a few miles west of Las Vegas, bristlecones range from 6,900 feet to 10,000 feet on Charleston Peak, edging the treeline near the summit. Yet they are the tallest plants in their environment. Routinely, trees seek immortality.

Some bristlecones are nearly 5,000 years old. Second place goes to the Alérce, a conifer related to the giant sequoias, with some specimens estimated to be around 4,000 years old. In your own backyard, the age of a living tree can be estimated by counting its whorls, or groups of branches. But there's always the catch of absent whorls as a result of limbs relinquished, and Southern pines grow more than one whorl a year. The age of the bristlecone pines was estimated by tree rings.

The wider the ring, the faster a tree grew that year, the more food it photosynthesized, the more water it moved—all demarked by a line that begins with new growth in the spring. During a drought or a freeze, the ring is slender, reflecting limited growth. Tropical lines are obscure, and even the rings of temperate trees can show false starts as a result of a cold snap. Within the ring for a single year, two areas of growth can be distinguished. The innermost part of the ring is called early wood to denote spring growth; late wood that grows in summer is denser and darker. In Mattheck's computer program, he compares additions to a blueprint to the outermost cambium layer in that it grows to improve upon history. D'Arcy Thompson had called the cambium layer a phase of nonequilibrium, a condition of flux, a key issue in the rhythmic phases of growth.

Microscopic details in a ring can contain evidence of ancient forest fires, volcanic eruptions, droughts, and floods. Climatic shifts that occurred thousands of years ago left clues that are studied by dendrochronologists, keen to measure the recovery time for previous global warmings. Clues include carbon isotopes. Since carbon molecules vary in weight, that variation in different rings can indicate warming trends. The global carbon budget is also studied in deep sea cores, which contain fossil plankton. Like trees, these single-celled plants rely on photosynthesis of a carbohydrate diet, and the ratios of carbon dioxide to water that photosynthesized can be discerned. Within the rings of foxtail pines and western junipers in the U.S., there is a record of a warming during the years 1100 to 1375. This research by Lisa Graumlich of the Laboratory of Tree Ring Research at the University of Arizona in Tucson matches clues in the literature of the era previously regarded as fiction: vineyards in England, farming in Greenland. Tree rings also contain a record of repairs by reaction wood.

The mechanics of self-repair remain mysterious. Auxin hormones seem to trigger reaction wood. These hormones are synthesized in the very tips of shoots, or apical tissues, but travel towards the cambium, the tissue for new growth in both trunk and limbs. The concentration of auxin is greater in lower parts of tree limbs and on one side of a stem or branch, creating a condition known as epinasty, which explains the shape of the branches for *Euphorbia robecheii* in Africa and many subtle curves in plants.

But what signals the auxin to act? Tom McMahon, who teaches biomechanics at Harvard, reckons the living cells might respond to ray initials, which extend from the cambium to the sapwood, a reservoir of pith cells. "The wood rays are in a position to sense the local bending caused by gravitational forces or the wind," he notes, because that bending "produces a squeezing of the wood on one side of the branch, and a stretch on the other side." He suggests that the wood rays could be "sensors" that tell the cambium where to speed up secondary growth.

To Thoreau, the whole tree was but a single leaf, and to Goethe, the parts of a plant developed from a single archetype, like a leaf. Both men were metaphorically but not mathematically on the verge of a hot concept.

Fractals denote a pattern obvious in trees, branching into smaller and smaller branches, just as our lungs and blood vessels do. Growth is by repetition, bifurcation, repetition, bifurcation. The form repeats itself into the minute—to infinity, it has been suggested, were a structure to exist so long and so large. The name "fractal" combines fraction and fracture—the creation of a new surface by breaking into fractions of the original. "Fractal" was coined by an IBM researcher and French mathematician, Benoit Mandelbrot, whose first book on the subject, in French, was published in 1975; his manner and presentation did not enhance acceptance. Mandelbrot's background was in economics—he had studied cotton prices—and when he ventured uncanny expertise in other disciplines, the response was somewhat territorial. The concept itself had great appeal, and like many in science, it seemed to be something in the air.

In 1975, the same year that Mandelbrot published his book, Tom McMahon wrote an article in *Scientific American* that began with exactly the same theme: "The branching form of trees are deeply pleasing to us and would seem to serve the needs of the trees themselves. In their proportions there is a harmony that makes us wonder if we could discover a principle for their mechanical design."

Such a principle could clarify similarities among the dissimilar. McMahon wrote, "For centuries observers of nature have wondered if common rules could be found that would apply to such fundamentally different kinds of ramifications as branching in trees, in watershed drainage systems and in anatomical structures such as blood vessels and bronchial tubes." Various formulas for the branching angle and the dimension of branches had been proposed. McMahon drew upon the work of a geomorphologist, Arthur N. Strahler, who studied the branching patterns of river tributaries, as did Mandelbrot.

Strahler's mathematical view produced a logarithmic chart of incremental progression. It began with a lower ratio of 1 to 2, which is what you would have to begin with to fork. Yet Strahler found no theoretical upper limit. This ad infinitum potential intrigued Mandelbrot, who applied a geometrical pattern. Triangles of repetition and their parallels in nature led to comparisons to coastlines, and speculations that fractal patterns were not just common but universal, and encoded in DNA. In the tributaries of the human body, the function of fractals has been known for a long time.

"Many problems of a hydrodynamical kind arise in connection with the flow of blood through the blood vessels," D'Arcy Thompson wrote, and as a solution, described how the branching angles contribute to the principle of minimal work, what Thompson called a "fundamental in physiology, and which some have deemed the very criterion of 'organisation.'"

To define minimal work, Thompson wrote what Mattheck and others now repeat: "a maximum of efficiency at a minimum of cost." Cost in an organic form is metabolic and material. The branching of veins and arteries into smaller and smaller capillaries allows for greater areas to be covered, with blood transporting oxygen to our fingertips and toes and within reach of every two or three cells of our body.

Various amazing statistics have been calculated from this, such as: our blood vessels, laid end to end, would stretch 10,000 miles. But the telling numbers denote surface to volume, a ratio of dimension and scale that dominates life on earth, and explains many forms. The total *surface area* for exchange in our blood capillaries is about one and a half acres.

The major clue of efficiency is the dimension of large to small. If a sedentary person begins to exercise regularly, the capillaries grow in number, and veins and arteries are forced to grow by the same proportion.

Just as we have blood pressure, trees have water pressure. Leaves pull water up by transpiring. As the water evaporates, the space is filled by pulling more water from the roots, which supply minimum pressure. When the water vessels of a tree are cut, there is a sudden sucking sound, of air being pulled into the system. For every gram of carbon fixed by a leaf, several hundred grams of water are transpired. The opening and closing of stomata, the pores of the leaves, controls the flow. At night, leaves relax, and the hydraulic lift ceases. This allows water to flow back into the soil, and plants growing around trees benefit from this during a drought, because the tree has collected water their roots cannot reach, and they share the plenty.

A tree absorbs carbon dioxide, and gives oxygen in return. Most of the gas exchange occurs in the leaves, which can number into the tens of thousands and bury a tennis court. A mature maple may have 20,000 branches. An acre of forest can have eight acres of leaf surface, three times as much as an agricultural crop. The vast surface area has the same purpose, exchange and diffusion, with the tiny stomata in the leaves absorbing carbon dioxide. The veins of leaves are tiny, like our capillaries.

Galileo was the first to suggest that trees cannot grow beyond a certain size else they would collapse from their own weight. Over 350 years ago he published "Dialogues Concerning Two New Sciences," written as a conversation with a student. "Thus it is impossible to build enormous ships, palaces or temples . . . nor could nature make trees of immeasurable size, because their branches would eventually fail of their own weight."

In 1881 a British mathematician named A.G. Greenhill used a formula to test this theory. The formula was the same one that D'Arcy Thompson used, devised by Archimedes: The diameter increases by the cubed power in ratio to the height squared. Greenhill used as his model the flagstaff in Kew Gardens, a pine from British Columbia that was 21 inches in diameter at the base, and 221 feet tall. He concluded that the tree could not have grown more than 300 feet tall. D'Arcy Thompson points out that this was the same figure that Galileo gave as the utmost limit of a tree. The giant sequoia is within the utmost range, with a height of 275 feet and a diameter of 26 feet.

Tom McMahon did calculations on all of the trees in The National Register and on the average found that the diameter was proportional to the $3/2$ power of its height. J.E. Gordon wrote about a tree 360 feet tall, 60 feet above the theoretical limit, and Claus Mattheck says a healthy tree should not collapse from its own weight no matter what the height, claiming that "the weight of the tree is a negligible loading case compared to bending by the wind."

McMahon thinks the elasticity of trees may be more important than their material strength and notes that each tree has a singular frequency. A tree without leaves makes "a fairly satisfactory tuning fork," says McMahon, who has a reputation for pushing trees. "The best way to get a tree into resonant vibration is to push it rhythmically, the way you would push a child in a swing, applying a little shove each time." Once the tree is oscillating on its own, he measures the time it takes for "a number of vibrations, which determines its natural frequency." McMahon suggests that this frequency is predictable according to the length of the trunk. The shorter the tree, the higher the frequency, and limbs are more elastic than trunks, which makes sense. The younger the tree, the more elastic. A sapling with a load of snow can spring back into shape.

If trees are so exemplary, why did Claus Mattheck find it necessary to study bones to enhance his Soft Kill Option? For starters, bones do not shed hazard beams.

Few of us think of our bones as plastic, but their malleability is evident to an orthopedic surgeon who monitors their repair. A badly set bone can actually grow to straighten itself, filling in the weak spot by increased mineralization, shrinking the unnecessary bulge by lack of minerals. The agent of change is calcium, the same material that strengthened the sand dollar test, and dissipates in the bones of astronauts. Bones designed to carry a human body against gravitational forces on earth quickly adapt to weightless conditions, a Soft Kill Option at work. When less work is required of a bone, it shrinks in size. This mechanism was what Mattheck wanted to recreate in human designs.

Getting rid of waste in designs dominated postwar industry in Japan, and the approach included refining the efficiency of the production line as well as the early stages of design, trimming materials and energy required. According to a December 1990 report in the *Wall Street Journal*, the U.S. created five times the waste that Japan does for every dollar of goods sold, and twice that of Germany. "The most waste of energy is in oversized and much too heavy structures," Mattheck wrote, and with this in mind, he turned to bones.

Changes in bone growth leave a record, with lines that resemble the blueprints drawn up by engineers, sometimes in a blatant form. Such was the case in 1866, in Zurich, when an engineer named Cullmann wandered into the dissecting room of an anatomist named Meyer. As D'Arcy Thompson recounts in *On Growth and Form*, Cullmann had been busy designing a crane. Meyer was studying a femur. Along the neck of the bone were visible, intersecting lines, known as trabeculae, a

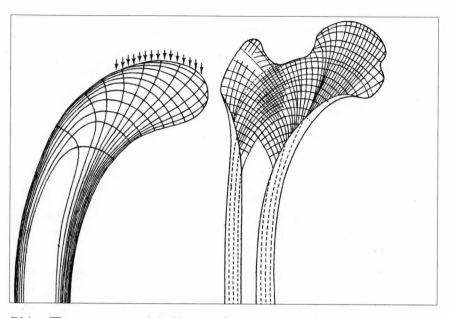

D'Arcy Thompson compared the blueprint for a crane drawn by an engineer named Cullmann, to the lines known as trabeculae in the head of a human femur on the right. Both have lines that express how loads are distributed.

result of the stress and load imposed on the bone, and a record of its responsive growth. The engineer is said to have taken one look and cried out, "That's my crane!"

The lines of stress were in response to the loads imposed by gravity. Other bones found different ways to resist gravity, and they figure in the wings of living cranes.

3. Wings

I said, Try to remember that nobody can do everything. He looked wry and said, "Except Leonardo da Vinci; did you know that on top of everything else he could do he was one of the great singers of his time, with a beautiful singing voice, and that he won first place in a competition at Milan over singers from all over Italy and won it accompanying himself on a new type of lyre invented by himself, made out of silver, in the shape of a horse's skull, which gave out powerful and ravishing sounds of a kind never heard before?"
— NORMAN RUSH, *Mating*

A lot of the dreamers had the right dreams but didn't know how to pursue them. Leonardo sketched and talked but he didn't do anything about it. We've gone a bit further.
— PAUL MACCREADY, *Zero, Three, Bravo*

AN "UNUSUAL quickness of eye" was once given credit for Leonardo da Vinci's study of birds in flight, which he undertook in 1505. In the first treatise on the subject, he ventured that "a bird is an instrument working according to mathematical law, which . . . is within the capacity of man to reproduce." About 400 years later, Orville Wright wrote: "Learning the secret of flight from a bird was a good deal like learning the secret of magic from a magician. Once you know the trick and know what to look for, you see things that you did not notice."

On more than one occasion Orville Wright was looking for lateral control. He drew on thousands of trials and errors, some with bugaboos of fatal proportions that, in the era just before Kitty Hawk, involved transitions from models to full scale. An invaluable assist on mistakes not to repeat was provided by the Wright Brothers' friend in Chicago, Octave Chanute, who pinpointed the problem as one of stability. Models were inherently stable, or rigid, but like a tree that needs to deflect drag, or columns for the Los Angeles freeway that should billow when the earth moves, stability amid the elements required a little instability in the structure. When a pilot needs to respond, so do the wings, which is how pilots of small planes come to think of their machines as extensions of their bodies, dipping their wings in salutation. The Wright Brothers already had a tail rudder as an essential configuration of gliders, but the nose would come later in the form of a prop described as an aerial screw.

The brothers' pivotal move was to experiment with warped wings, based on the way a turkey vulture reduces turbulence at low speeds. The tips of primary feathers can extend and flex like fingers, creating slots that induce drag. To deal with a shifting current, a soaring bird such as a vulture levels itself by twisting one set of wing tips down, the opposite up. This is positive warp, a function in aircraft now satisfied by ailerons hinged on the wings.

The story of pioneering feats in aviation has been done to death, you would think, but an aviatrix and one of my neighbors in Africa put the value of history this way: "When the first man was made, he wandered alone in the great forest and on the plains," Beryl Markham cited a Nandi legend; "And he worried very much because he could not remember yesterday and so he could not imagine tomorrow." Since flight is considered humankind's greatest technological achievement, the way tricks can be overlooked says much about any attempt to draw blueprints from nature. Three points emerge of insight, oversight, and hindsight. The first: In an age of space shuttles and the supersonic, the aeronautical engineers who are making vital contributions continue to draw ideas from birds. Two, the jury is out on whether Leonardo accurately perceived the downwash of pigeons' wings or if his ornithopter could ever reach an altitude higher than a museum ceiling. Finally, the development of aviation has strong parallels with the evolution of flight in nature, so when seeking other blueprints, understanding the evolutionary history may shorten the R & D phase.

The human-powered plane known as the Gossamer Condor involved the essence of evolution, a novel rearrangement of available parts. After a successful flight of the Condor in 1977, Paul MacCready, who came up with the idea, said it could have been done fifty years earlier. In fact it could have been done in 1903 at Kitty Hawk. Certainly the bicycle parts were available, and so was the piano wire. The Mylar that covered the wings was new, but as MacCready noted, silk would have sufficed.

A couple of soaring birds (a hawk and a turkey vulture) inspired MacCready, such an avid observer of birds that he nearly became one. As a three-time U.S. national soaring champion, he was said by his competition to fly like a god, to evoke the awe felt by ancient cultures that considered wings supernatural. "It's a wonder more things don't have wings," MacCready once mused, "because wings give you freedom in a third dimension, to escape predators, forage over huge areas, and migrate quickly to better food supplies." Insects outnumber us, outsmart us, and as a group are considered offensive until someone whispers "butterfly." This is the grace that MacCready brings to aviation, a simplicity that did not escape Leonardo, who entered in his notebooks, "Given the cause, nature produces the effect in the briefest possible way."

"Humans drew inspiration from birds as aviation was getting going," MacCready reflected, but people "continued developments on their own, forgetting about birds—unless they worried about ingesting them in their engines. *After* developing retractable landing gear, flaps, navigation devices, autopilots, and active controls to provide stability for unstable vehicles, people realized birds had been doing these things for 100 million years."

An aeronautical engineer who studied at Caltech, MacCready built model planes as a boy. At first, he constructed them from kits, then he designed his own, which have been described as unorthodox. At the age of 14, one of his designs broke the international record for endurance. It flew for thirteen minutes, courtesy of a rubber band. Thirteen minutes is an eternity in experimental flight; the Wright Brothers made history in a mere twelve seconds. During the sixties, MacCready spent much of his time seeking what most pilots avoid, investigating turbulence by flying a single engine into thunderheads, seeking the parts of clouds where rain and hail are generated.

The Gossamer Condor looked as though one of the kids in *E.T.* had flown his bicycle through Saran Wrap. It weighed 70 pounds. The wings were enormous, spanning 96 feet, longer than those of a DC 9. The Condor could cast a shadow over the Wright Brothers' biplane, spanning 22 feet, a comparison made easy since they hang alongside each other at the National Air and Space Museum in Washington, D.C.

Now 70, MacCready is the founder of AeroVironment, Inc., based in Monrovia, California. His inventiveness is mainstream enough to have earned him a feature story in *Reader's Digest*. AeroVironment thrived in 1990, with a revenue of $17 million. General Motors bought 15% of the company and funded MacCready's team to develop an electrical car known as the Impact. Like the SFB and SEB groups described in Chapter One, the bent of his organization is to draw lessons in economics from nature, and some designs (such as efficient wind mills and solar powered cars) tap the elements directly. The interdisciplinary approach is so inherent that many on staff are hybrids; Peter Lissaman, who refined the airfoils for the Gossamer Condor with computer models, also published a paper on the V formation of geese. Martin Cowley, who works as a designer on cars and aircraft, is an architect who deals with engineering on such a regular basis that when I asked if he was an engineer, there was a two-second pause before he replied, "I do engineering, but . . ." Equally unconfined are projects, from pollution monitors to piecing together a flying machine based on Leonardo's drawings. The Impact was undertaken after considerable success with solar-powered cars. The wheels for the Impact have tires that reduce drag, to such an extent that were they on a normal, gasoline-powered car, the car's fuel needs would be cut to one-third of its present requirements. "The Impact exploited every technology to reduce drag," MacCready says, "tires, streamlining, bearings, weight, etc." Energy usually dissipated when applying the brakes recharges the batteries.

In one project, the applied science was natural history. In 1984, Mac-Cready decided to build a flying replica of a pterodactyl, a Barneybird that lived over 65 million years ago. The design was based on fossils found in Texas. The inclination towards authenticity was such that Mac-Cready organized a workshop with paleontologists to review details of organic structure. It was decided beforehand not to cheat with modern aeronautical advantages, such as ailerons or flaps. Among the details paleontologists imparted was that the pterodactyl had had fur on its body. Since the goal was to make the replica "biologically accurate," fur was duly noted. Also, there was no tail, but the huge head was positioned far ahead of the wings. "Obviously it would be unstable," MacCready said; "like an arrow shot with the feathered tip in front. But birds and bats are also unstable. They are able to fly by using active control. The replica had a brain—an autopilot—to provide stability." The engineers allowed themselves the luxury of wheels. The test, after all, was in the air, since the goal was nothing short of the first flapping aircraft.

There was hardly the promise of a large order from British Airways, and the exercise seemed "the height of eccentricity," but what Mac-Cready hoped to do was to expose some parallels in the evolution of natural flight and human flight.

For example, among the flappers of the medieval, there were brave leaps with inadequate equipment. Winglets is the term evolutionary biologists apply to the early extensions of insects, which may have drifted on wind currents the way seeds disperse. Then came a series of short glides, greater wing span, longer glides, better control, and finally propulsion. That's a quick take of about 350 million years for nature, 500 years for human efforts. In both cases, certain constraints configure, which is why everyone loves Dumbo the Flying Elephant.

The constraints on propulsion are the most intriguing, because economy is an issue. Both birds and aircraft expend more energy on takeoff than in level flight. Vultures and hawks glide to conserve their metabolic fuel, and pilots conserve fuel by using the jet stream, a boost of wind currents that travel up to 360 mph. Depending on a single source of propulsion for all aircraft is a constraint. It is a constraint that we can anticipate becoming more constraining. Major oil sources are foreign, and aviation fuel is not cheaper than bottled water. When scientists began to look at the construction morphology of natural flight, they found themselves substituting the words "opportunity" and "possibility" for "constraint."

"Physics put some constraints on the system," says Steve Vogel of Duke University; "What we see is only a tiny subset of conceivable creatures." Some organic constraints have to do with available material which we can overcome. There is no need for such displays as peacock feathers on a 747. D'Arcy Thompson alluded to a wider possibility when he described his study of organic form as only a small part of the "forms assumed by matter under all aspects and conditions, and, in a still wider sense, with forms that are theoretically imaginable." The Gossamer Condor was not a blueprint of a hawk, a vulture, or even the hybrid its name implies. It was theoretically conceived, with inspiration from natural forms.

The evolution of so many forms on the wing allowed flying creatures to tap a wide variety of fuel sources. If one source failed, alternatives were developed, by developing ecological niches, and great migrations. Their maneuvers vary: a hummingbird hovers to gather nectar; a fish eagle uses talons to pluck a meal out of a lake; and some birds bound between strokes, putting their wings at their side. This diversity led to different wing structures, composed of different materials in different configurations. Two-thirds of all creatures on earth fly. A disproportionate diversity can be claimed by insects, which may have gone back to the drawing board, to increase maneuverability in order to avoid becoming a meal when the first birds began to fly.

Because insects were the first to fly, their biomechanics are the most sophisticated, the most baffling, and the least understood. As J.B.S. Haldane wrote in *Possible Worlds*, "My own suspicion is that the universe is not only queerer than we suppose, but queerer than we *can* suppose." Even with high speed photography and frame-by-frame analysis, the flight of a bumble bee appears to defy the laws of aerodynamics, but as Paul MacCready put it, "Aerodynamic understanding is perfectly consistent with the flight of gnats, bees, bats and 747s." Leonardo's views are defied and alternately confirmed by the freeze frame. This may be because a pigeon can do many things with its wings at different speeds, as if it were shifting gears.

That a fly has a three-speed gear box to control its wing beat is mind boggling and perfectly logical, and, like the many geared inventions that Leonardo sketched, hidden from view for too long. The fly's gear box (pictured below) was published in 1991, in a book from the SFB project on Construction Morphology. For some reason, Leonardo's notebooks went unnoticed until the 1800s, languishing like stacks of the *New Yorker*; and, when finally exposed, his blueprints for flying machines were reproduced only in two dimensions, scaling heights as wall posters. For all their notoriety, not one of his ornithopters got off the ground, but a glider based partly on his drawings provided a means of escape for Bruce Willis in *Hudson Hawk*. It would take an unusual quickness of eye to discern that this flying machine merely glided; the illusion of flapping was created by the camera angle, which was low. "Da Vinci's plan had no device to counteract torque, so the person would have spun faster than

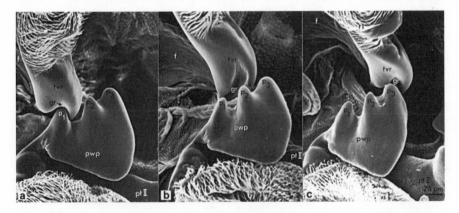

An SEM photograph of the three-speed gear shift for a fly's wing, allowing the fly to change the wing beat frequency. First gear is on the left.

the prop," MacCready said. Had the AeroVironment engineers stuck to the original drawings, "It wouldn't have gotten off the ground."

◆ ◆ ◆

HOW clearly the Italian artist saw things is rarely questioned, for he honed the most distant point of perspective (when perspective itself was new in art) and apprenticed with a sculptor in Florence known as Verrochio ("True-Eye"). His paintings are distinguished by his use of light, and its reception. "All our knowledge," he wrote, "finds its origins in our perceptions." So does our foolishness, and with his *Mona Lisa*, landscapes on the right and left sides of *La Gioconda*'s head do not match, as if she obscured a geological fault the size of the Great Rift Valley. His anatomical drawings of humans featured in the first medical textbook, and his maps of bones in a bird's wing were the first so detailed.

Aristotle before him had reckoned that air perpetuated flight; he thought an arrow, for example, was pushed by the vacuum being filled behind it. Yet Leonardo rightly saw flight as overcoming resistance and nearly scooped Isaac Newton's third law of motion by his recognition that "the movement of the wing against the air is as great as that of the air against the wing." (Newton wrote, for every action, there is equal and opposite action.) Leonardo saw that air had growth and form; it changed in density, it had velocity, it developed turbulence. Air moved the ocean as it moved clouds. It played with smoke rings. Air disliked vacuums and voids. The atmosphere might be invisible, but it was alive with pressures. Leonardo saw patterns in the behavior of air that were similar to those of water when it confronts a body.

His study of the way fish move in water led to a proposal for a spindle-shaped hull that created less friction and greater stability than round-bottomed boats. It resembles some hulls now used in the America's Cup race, and in a cross section, a fish tail. Beyond the sleek, he also designed the first double-hull, the first submarine, first car, the first lock and dam system, the first urban plan. There were so many firsts that it is remarkable that he found time to render *The Last Supper.* In the ripples that form on the surface of water, he noted the transmission of energy that today is known as wave theory, and he understood that the peaks and troughs that occur in the ocean are the same in pattern as the vibrations of stringed instruments, such as a lyre. Controlling the number of vibrations creates overtones.

Leonardo felt that if you couldn't apply numbers to a situation, there was "no certainty." This, among other things, endeared him to D'Arcy Thompson, who mentions Leonardo on page one of *On Growth and Form*. His regard for precise measurements was such that he invented the first pedometer, a mileage gauge to mark the revolutions of a cart wheel, and a contraption to test the tensile strength of wires. Sand was poured into a little basket supported by the wire until it reached the breaking point. Then the basket of sand was weighed.

Some of the sketches from his notebooks were so exquisite as to be selected for the world's most enviable art collection, belonging to Queen Elizabeth II. In 1980, Armand Hammer, the former chairman of Occidental Petroleum, bought the last set of Leonardo's notebooks in private hands, for $5.6 million, then a record for an autographed manuscript. The so-called Codex Hammer, which covers Leonardo's notes and drawings from 1506 to 1508 only, was expected to bring closer to $10 million when auctioned at Christie's in New York in 1994.

Twenty-five of the most intriguing technical and scientific drawings were seized upon by IBM in 1951 to assemble three-dimensional models of plans both practical and theoretical, including a spring-driven car (drawn long before watchmakers used springs), a parachute shaped like a triangle, and among the flying machines, a hang glider. The IBM exhibit toured galleries and universities throughout the U.S., "intended," as its literature proclaimed, "as a spur to creative curiosity—the quality that leads man to ask questions about the physical world around him, to dream dreams for its progress, and then to search for the answers that will turn the dreams into reality."

One of Leonardo's flying machines was an attempt at a helicopter, said to be partly inspired by the way vines ascend in coils. This aerial screw had a single massive rotor, a spinning top of linen that was to be powered by four people running in a circle. It does not look as if it were capable of flight, but then few helicopters do. An airplane propeller is also known as an aerial screw but bears little resemblance; aerial screws have been around. Archimedes invented them for raising water, and they were used in Egypt for irrigation. The true rotor is a mathematical quantity having magnitude, direction, and position, and the radial engines of airplanes rotate around a stationary crankshaft. Leonardo got into rotations of all sorts, including the arrangement of leaves and petals and branches known as the Fibonacci sequence, reinvented by his fellow Italian Leonardo Fibonacci in 1202.

According to Martin Cowley, who worked on the ornithopter design for Hudson Hawk, Leonardo had no single unified blueprint for this notorious flapper. "We went through all of the sketches," he said, "but there never was a single page with a full blueprint. He had one idea for pedals, another idea for wings, another idea how the pilot was to be strapped in. What we did was, as engineers, we went through the sketch-book and pieced together the most logical combination of pieces to make a flying machine." It was like dealing with a jigsaw puzzle, and by necessity, they "took some artistic license." While Cowley found some of Leonardo's ideas practical from a mechanical point of view, the budget for the movie prop, such as it was, did not allow AeroVironment to fully pursue the notion of a human powered flapper. "It's conceivable," MacCready allows, "but a hard way to do an easy job."

Like Leonardo, British photographer Stephen Dalton plays ravishing music, on a harpsichord he built himself, but he is best known for his high-speed photographs of creatures on the wing. In his book, *The Miracle of Flight*, Dalton estimates that the ornithopter would require gargantuan human breast muscles, 6 feet across, to propel it. "It is a pity that in all his works Leonardo clung so tenaciously to the concept of the flapping wing," wrote Dalton, who favors gliders as a pilot himself, but found only one (the same one featured in the IBM exhibit) among the many drawings. Dalton reckons that by trying to reproduce the flapping wing, Leonardo's "genius was misguided."

Leonardo compared the flapping wings of birds with the back stroke of a swimmer, and the circular pattern of oars, dipping backwards, when in reality the breaststroke would have been more accurate. "Such a conclusion seems hardly surprising considering the speed at which birds vibrate their wings," Dalton notes. "Even the slower actions of eagles and gulls were too fast for accurate analysis, which was only made possible by photography." Others credit Leonardo with acute vision. "He observed and recorded in his drawings the complex sequence of pigeons' wings fluttering in flight," wrote Leonard Slain in *Art and Physics*. "It was not until time-lapse photography was invented 300 years after he worked that anyone could slow down these visual blurs, and then the studies photographers made confirmed what Leonardo had seen."

The Italian artist believed support came from below, which is partly true for the flapping wing but not for fixed-wing aircraft. Generally, it is the flow of air across the top of a wing that creates lift; the pressure from below decreases. In both birds and aircraft, the top of a wing has a greater curve, or camber, while the bottom is flatter. The faster flow over the top creates less pressure. The concept of lift gains clarity by the very term, which denotes something being raised from above.

Leonardo imagined the flapping wings of pigeons condensed the air below, producing an instant skyway of support. Butterflies and moths create vortices of air beneath their wings by twisting and turning on the downstroke. It was (and is) difficult to distinguish the forces of lift from propulsion, since natural wings work at both. Because lift is maintained when wings are in a fixed position, soaring and gliding birds imparted the basic blueprint. So the fixed wing, not the flapper, advanced aviation, and it may have done so because it was all people could accurately perceive and because it worked.

Some of the right stuff was discovered a century before the Wright stuff, when a British observer, Sir George Cayley, figured out that flapping was a source of propulsion, not lift. Cayley was able to measure one of the mathematical laws that Leonardo predicted, calculating that a gliding rook was capable of maintaining altitude when its airspeed reached 25 mph. The basic formulas of flight include weight versus lift, and thrust versus drag. Cayley also saw that birds require greater energy for takeoff than for straight flight, and he turned his attention to soaring and gliding and the fixed wing. Some historians, including Stephen Dalton, credit Cayley as the builder of the first successful airplane, a glider that flew across a Yorkshire valley, piloted by his "reluctant coachman." Cayley's findings were first published in England in 1809, but it wasn't until 1866 that his ideas gained support, from another observer of birds.

Francis Wenham discerned the value of a cambered wing, and found that a long, narrow wing provided greater lift than a short, broad one. A Frenchman was quick to follow, with his report on soaring birds. In 1881, Louis Pierre Mouillard published *L'Empire de l'Air* which in turn inspired the German engineer Otto Lilienthal. Lilienthal studied the way a bird uses its primary feathers to create twisting slots at the wing tips to design and fly beautiful biplane gliders that influenced the two bicycle mechanics from Dayton, Ohio.

The first airline was founded in 1909, only a year after Wilbur Wright gave a demonstration flight to a stunned crowd in France. The airline consisted of zeppelins, named after the German count who designed these dirigibles, which flew thousands of passengers on regularly scheduled flights within Germany, and from Frankfurt to New York and Rio. With a top speed of 75 mph, airships crossed the Atlantic in two-and-a-half days, and passengers enjoyed the luxuries of an ocean liner. Dramatic newsreel footage of the explosion of the *Hindenburg* in 1937 put an end to airship passenger travel, for reasons as unfortunate as the crash. The Germans were forced to use explosive hydrogen, while nonexplosive helium shipments sat on the dock in New York Harbor as restricted military supplies; zeppelins had bombed London during World War I with terrifying stealth.

While sabotage and lightning were entertained as causes for the explosion of the *Hindenburg*, the latest theory involves a phenomenon known as Saint Elmo's fire. A ghostly, blue, coronal discharge that can loom on a ship's mast during a storm, it was thought by Mediterranean sailors to be a sign of their patron saint, Elmo, an Italian bishop of the fourth century. Saint Elmo's fire is accompanied by cracking and hissing, and results when electrical charges in the atmosphere differ from those of an object, such as a ship's mast or part of an aircraft. It commonly occurs on airplane wings and props during a thunderstorm, and when aircraft fly near a sandstorm or cumulonimbus clouds. This proves uneventful because of the position of the fuel and its containment.

Hydrogen-powered cars are being developed, and the U.S. space shuttle employs bottled hydrogen on board for various uses. Airships deserve reinvention, and John McPhee chronicled an attempt during the early seventies in the form of a hybrid officially called the Aeron 26 but remembered as the Deltoid Pumpkin Seed, the title of McPhee's account. A decade later, I saw it sitting in a hangar near Princeton, New Jersey, gathering dust. "Any form of aviation must be experimented with," McPhee was told by a Navy veteran of airships, Everett Linkenhoker, who turned to the Aeron 26 project after bitter disappointments with the Navy's status quo approach to lighter-than-air craft. Said Linkenhoker: "All through the Second World War they should have been experimenting, but they were flying the same type of airship up till the very last. . . . We were down to twenty ships by 1953. Nothing new was cranked into the picture; it just had to reach its end." The Aeron 26 completed a number of successful maneuvers, even though it had no wings.

The Wright Brothers built a glider with a 12 horsepower engine. They made a thousand glides before they added the engine, which weighed 200 pounds, and which they virtually invented. Some of the lightweight features of their design came from their experience manufacturing bicycles. A bicycle wheel, like an umbrella, is a contained tensile structure; the spokes are in tension, the compression is the air in the tires. Consequently, the weight of the rider hangs in suspension. For the energy required, a bicycle is rated the most efficient form of locomotion for humans, as mentioned. The leg muscles that power it are strong, more developed in humans than breast muscles. When human-powered flight was finally achieved, the power was provided by leg muscles. In the case of the Gossamer Condor, the legs belonged to a bicycle racer.

Several aircraft, notably British and Japanese, had managed sustained human-powered flight, but none had satisfied the performance rules required to win the Kremer prize, established by a British aviation buff named Henry Kremer in 1959. To win the award, no incline could be used for launching, no lighter-than-air gas could assist, and no ballast could be jettisoned. The ground had to be level and flight attained with winds below 11.5 miles per hour. The craft had to fly just over a mile and complete two 180° turns, banking in a figure eight around pylons set half a mile apart, and, finally, it had to clear a 10-foot hurdle at both ends of the course. The prize was £50,000—about $87,000 at the time, and it was the prize money that provided the initial motivation for Paul MacCready. "Mentally, I tried several conventional designs, but found they couldn't succeed," he recalls.

In 1976, while driving on a vacation with his family in New Mexico, he saw a hawk circling overhead, and he timed its revolutions with his wristwatch. Then he did the same for a turkey vulture and later made some calculations on the back of an envelope. "Working on scaling laws to quantify bird circling flight and sailplane circling flight provided the 'lateral thinking' for the solution." The human engine is equal to about half a horsepower. To compensate for the limited power of a human, he would triple the lift value. This required a vast set of wings, exceptionally lightweight.

Rather than wooden struts, MacCready designed wings reinforced with piano wire. The wheels were plastic. Diminishing weight was everything. The pilot, Bryan Allan, weighed only 135 pounds. He took off his wristwatch and a turquoise ring before climbing into the cockpit, which was also transparent and resembled a ziplock bag. Allan pedaled for just over a mile at an average speed of 10 mph in variable winds.

If the ornithopters that Leonardo envisioned require human breast muscles of great proportions, birds and bats have developed towards these dimensions. Insects employ a different kind of fibrillar muscle to power their wings, which is why you hear them buzzing. The strength of these muscles can be measured, along with oxygen consumption. Many measurements have been tortured to try to explain the bumblebee's flight; its fuselage seems too large to be supported by such tiny wings.

The pterodactyl may have been pushing the envelope when it comes to wing loading. While some of these ancient flying animals were small, even bat sized, the fossils from Texas suggested a wingspan of 36 feet. Pterosaurs were previously thought to be gliders, but, according to Martin Cowley, the designer at AeroVironment, pterodactyls would have to have had some flapping ability because, Cowley explained, "if they only glided, they would have no way of gaining altitude." And even though the bones of the wing were thin, as mass increases, the energy required for thrust increases in a predictable formula. A sea gull is twice the size of a robin, for example, and the gull's requirements for metabolic energy increase by a ratio of 1.4 to 1. No living bird has a wingspan even a third as long as that of the pterodactyl from Texas. The California condor has a wingspan of 9 feet; the Kori bustard of Africa, said to be the heaviest flyer today, weighs 35 pounds. The estimated weight for the pterodactyl is around 280 pounds.

Fossil evidence from various geological horizons suggests that pterodactyls persisted for over 100 million years, and met their demise alongside the dinosaurs as a result of a global event having nothing to do with their flight capabilities. But the 36-foot wing span makes the pterodactyl one of the largest known creatures ever to fly. The wings were similar to bats' in that they had a stretch of skin over their bones, rather than feathers. The pterodactyl gets its name from the fourth "dactyl," or digit, that grew into the main wing bone. Meanwhile, the long reptilian tail bone was apparently diminishing. The team from AeroVironment dubbed it the "Q.N." after its scientific name, *Quetzalcoatlus northropi*. ("Quetzalcoatlus" is a reference to an Aztec serpent god, and the species name honors John Northrop, the American aviation pioneer who designed a tailless aircraft.) So the Q.N. was theoretically imaginable.

To make it a possibility, the aircraft would be remote controlled, with computer chips in charge of wing warping. Assuming the function of the alula feathers were three small claws, located on the leading edge of the wing.

◆ ◆ ◆

"WHAT can be more curious," Darwin asked, than a human hand "formed for grasping, that of a mole for digging, the leg of a horse, the paddle of a porpoise, and the wing of a bat should all be constructed on the same pattern and should include similar bones in the same relative positions?"

The tube feet of a sand dollar come to mind, also a radiation of five. The bones that Darwin listed occur among mammals, a small group, after all, and so new to life on earth we should carry green cards. Disparate were the sources for wings—the reptilian, the millipedec, the botanical—yet they are all similar in form because of their function. The wings of samara, transporting seeds of ash, maple, and elm, are remarkably like those of dragonflies. While the pterodactyl grew a wing strut from its little finger, and birds from their middle finger, the finger bones of a bat so resemble our own that the name given their order (*Chiroptera*) means hand wing.

There are 700,000 species of insects, 9,000 species of birds, and a mere 4,000 species of mammals, 900 of which are bats. Variety seems to favor things with wings, and fish make you wonder, some with pectoral fins nearly as long as their body and capable of gliding in the air for 50 yards. Manta rays are capable of being airborne in a literal sense, giving birth during flight above the surface. Petrels, gannets, dippers, guillemot, steamer ducks, and auks fly submerged.

Bats can glide, but since the night air holds few natural thermals, they rarely find the means to soar. (The terms are often confused; gliding occurs in steady wind currents, with a predictable decline in altitude. Soaring relies on warm currents, rising. You can glide into a landing, but it would be a feat to soar into one.) Small bats are strong flyers, with a spectacular rating among researchers like Rick Adams and Scott Pederson, who dubbed them acrobats. The young zoologists screened the exit of a Wyoming colony, and watched small brown bats slip through inch-wide partitions. It was better than MTV. Some flew wing over wing, and others bounded, collapsing their wings to emerge like pinballs. Bats don't have the ecological range of birds, but distance does not intimidate them. After strong westerly breezes, North American species turn up on oil rigs in the North Sea.

Adams and Pederson described bats as the only true flying mammals in an article in *Natural History*, which featured a photograph of a flying squirrel in the same issue. Some squirrels glide courtesy of stretched skin held taunt in a fixed position. Granted, squirrels gain altitude by climbing trees, but glider planes have no inherent means of gaining altitude. If the test is lift, these squirrels have an airfoil that qualifies, and the giant red squirrel of Borneo can glide for 300 feet. With power, the Wright Brothers first managed 120 feet.

These squirrels deserve their fifteen minutes of fame, because their mode could demonstrate a first stage in the evolution of natural flight. There are two theories about the origins of birds on the wing. (A third theory on fluttering has few takers.) Theory One has ancestral birds running on the ground, making occasional leaps, spreading their winglets to extend the leap with a little glide. The second theory is that they moved from tree to tree by gliding. Squirrels demonstrate the latter. They begin with a steep dive, gain speed, level off a bit, and just before reaching the next tree, veer up to create a stall to reduce their airspeed. They can also execute 180 degree turns, maneuver around obstacles, and collapse their skin flaps by putting their legs together. Their airspeed is 14 yards a second. But when it comes to defining flight, the criteria of an airfoil would include the human body when a skier flies off the end of the ramp, and a sail full of wind.

In Jeremy Rayner's view, squirrels aren't true flyers; they're just going through a phase. Rayner, a biologist at the University of Bristol in England, considers gliding an adaptive stage. ("I have no doubt that bats evolved through gliding," he wrote. "On balance I consider this the far more likely model for birds.") He credits Darwin, who thought early bats looked and behaved like the giant red squirrels. While gliding squirrels in Asia are closely related to some bats, no fossil links have been found.

"You get fossil bats from Wyoming which are early Eocene," Rayner told me, "but they're already perfect bats. What we never find in the fossil record is half a bat. You can look at the archeopteryx, and say it's got flying feathers, it's got aerodynamic wings, so it's half between a dinosaur and a bird. But we've never found something that's halfway between a bat and a tree shrew, half a lemur—what we might expect to find. It didn't exist. Or it's not been preserved or it's not been found."

While bats are newcomers, emerging a mere 50 million years ago, archeopteryx lived 150 million years ago, in the real Jurassic Park, at the same time as the pterodactyl. Archeopteryx was so perfect a missing link that charges of fakery were inevitable. The first specimen was discovered in Bavaria in 1861, at a site known as Solnhofen. Nothing seemed as implausible as such a delicate form as a hummingbird evolving from a dinosaur, even a small version such as the thecodonts. But this missing link had a wishbone, or fused collar bone, in addition to wings and fossilized feathers.

The original function of feathers may have been insulation; the blood of birds is much warmer than ours. The average body temperature can be over 110 degrees Fahrenheit. Numerous specimens of archeopteryx have since been unearthed, and the evidence of more feathers fueled the theory that dinosaurs were warm blooded. By studying the wing structure, Rayner came to the conclusion that archeopteryx was basically a glider, with an ability to flap its wings for faster flight. That is not a compliment. Slower flapping is a more complex movement, requiring greater muscle control, especially during takeoff and landing. Rayner reckons lateral control at slow speeds was a problem.

Rayner defines flight as the ability to produce useful aerodynamic forces by flapping the wings. This sounds as if the Concorde represents a primitive stage, but his definition was meant to return the study of natural flight to a level of inhibition. "[In] all the earlier work done in flapping flight," he says, "people used models that kept the wing outstretched, changing the angle of instance of the wing, rather like a fish tail." In addition to a greater repertoire moving from upstroke to downstroke, flapping flight produces unsteady effects.

Born in London in 1953, Rayner was influenced by the discoveries of a professor at Cambridge named Torkel Weis-Fogh, who found that wasps moved their wings in a way that could not be explained by conventional means. Wasps appear to touch or "clap" their wings at the extreme of an upstroke. On the downstroke, the wings twist, or "fling." He called it the "clap-fling." None of this created a predictable wake, such as you see with the straight vapor trails of aircraft. Either these insects weren't flying, or biologists had to go beyond the aerodynamics for fixed wings.

Rayner was among several biologists to return to Leonardo da Vinci's stance. To understand flight, he needed to figure out how the air behaved. Four hundred and seventy-four years later, the focus returned to the downwash of pigeon wings.

As a student at Cambridge, Rayner studied math and became interested in fluid mechanics. He thought of working in meteorology or oceanography, but a Ph.D. project on birds and flapping flight led him to the University of Bristol, where he currently researches bats, photographing their movements on the wing, traced by helium bubbles that make for spectacular images. He is careful to explain that this research was experimental and that his previous work was theoretical. This turned out to be a modest reference to a theory he proposed at the age of 25.

His idea involved vortex rings, which behave like smoke rings. A round, central puff of air creates a ring that rotates. When geese fly in a V formation, they share vortices at their wing tips, increasing their flying time by reducing their work load. The clap-fling motion of wasps also produces vortices, a method of lift for moths, lacewing flies, and butterflies. Rayner suggested the downstroke of a bird wing produced lift with these vortices but the upstroke was passive. If this were true, it should be visible if the wakes could be traced. Rayner published his theory as his doctoral thesis in the SEB journal in 1979. The theory was proved a few weeks later.

Most scientists consider things to be moving at breakneck speed if their theories are tested within a decade. It happened that a researcher in Moscow named Nikolai Kokshaysky was working on the same problem, in the experimental mode. Kokshaysky photographed finches flying through clouds of sawdust. The swirls of sawdust showed tight vortices on the downstroke, and the photos ran in Britain's prestigious journal of science, *Nature*. "Very lucky," Rayner allowed; "It fit in perfectly." For finches, the gist was there but the images were not always sharp, and improving the clarity, and testing other flappers, became a goal at Bristol. In a Victorian building on Woodland Road that houses the zoology department, experiments in flight have a reputation for breaking boundaries and lending new clarity. When the first wind tunnel was assembled, a flip of the switch created gusts of 45 mph in the stairwell, blasting out windows thick as an ale glass.

I met up with Jeremy Rayner at the Bristol campus on a day of explicable calm. It was Friday before Easter break, and the zoology department was deserted except for the young professor and his English spaniel, Roland. Roland's owner is blond, clean shaven, and wears khakis and a dark jacket—a nautical neatness reflected less in his office, where mounds of papers are surveyed by a pterodactyl model suspended from the ceiling. Rayner was keen to fit the extinct flyer into a comprehensive study of bird and bat wing ratios, and his own theories about gliding in the evolution of flight.

Bristol's reputation for research include the innovative approaches of Colin Pennycuick, who studied gliding birds in the Serengeti by flying alongside them in his motor glider, switching off his engine. He came to think of the vultures and pelicans that he followed as doing the same; their flight was like aircraft with the engine switched off. Birds capable of flapping could switch on the engine, while a pure glider, like the squirrel, had a shallow angle of descent, and moves courtesy of passive lift. In addition to his field studies in Africa, Pennycuick led a team at Bristol (including Rayner) that came up with the innovation of helium bubbles for tracing wakes. The bubble itself is ordinary soap and water. The gas inside is a mixture of helium and air, making the bubbles neutrally buoyant. The method of tracing air patterns employs the same principle used by aerospace engineers, but helium bubbles are not that common. "Since we've done it, several helicopter manufacturers have picked it up," Rayner says. "Birds fly through it quite easily."

Rayner speaks of the geometry of the wingbeat, changes with flexes of the wrist, fingers coming together, the thing that separates a true flyer from passive lift, and says, "I can show you that visually." I pick up my tape recorder and notebook, expecting to move down the hall to a lab, having sped past Stonehenge with visions of live bats moving helium bubbles. Rayner boots up the computer on his desk.

The outline of a pigeon appears on screen. It begins to flap, gracefully. Still photos from the experiments have been converted to computer animation. "We only use stills, because with cine, you need high-speed film. To get the image, you need a lot of light, and with a lot of light, the birds see the bubbles and won't fly through them. So we photograph them in stereo. From two photos we use a machine called the stereo comparator; you see the two images side by side, and use the human eye to fuse them. My students do that part. With 3-D, the velocity of the bubbles can be calculated."

The vortex patterns resemble a gentle snowstorm, when flakes swirl and bunch in eddies around a big object. "Where you see the dots in the pattern, there's very little movement. Where you see a streak, there's rapid, intense movement. You have an impression of curvature. Now," Rayner explains, "if you had a picture of an aircraft, you'd have a straight line behind it, classic aircraft wake. But here you have a very strong downflow—effectively Newton's third law, What goes down must go up. The bird's pushed air down, and that provides vertical force. That's the lift propelling the bird. That's what I predicted theoretically, and what Kokshaysky showed in 1979."

But what worked for pigeons and finches didn't work for kestrels. "We got a completely different picture, with no obvious rotation." It was a surprise, and pigeons showed a range that he compares to the gaits of horses. "There's a lot of ways they could move their legs and still be stable and move forward. In the end a horse will only use one pattern at one speed, with few exceptions. This seems to be the same sort of phenomenon."

Bats have eight or nine muscles that drive the wing beat, birds only a couple. Yet the way the air behaves is virtually the same. "The big difference we've seen is in size, very little bats and little birds. It's hard to get little birds to fly in the lab to compare. We've done it with swifts, but that's about all we've got.

"The problem with bats is that with echo location, they sense the bubbles and think they're flying through fog. They'll do it two or three times, then they stop because they don't like it." Rayner rarely keeps bats in the lab for more than two or three days. "For conservation reasons, I'm reluctant to remove a wild bat from its natural habitat." But he did not reject a guilt-free windfall of data, which occurred when someone in the area cut down a tree in winter. "It was full of hibernating bats. They woke up, but they couldn't be released into the wild; it was the wrong season."

Bats' foraging techniques are described as aerial hawking, skycatching, trawling, and gleaning. Aerial hawking involves cruising; skycatching means detecting insects from a perch. Trawling is fishing for insects on a water surface, and gleaning means to find a sitting meal—on a leaf, for example. All rely on echo location, an ultrasound from the larynx tuned by FM.

The element of surprise is unnecessary for megabats, which eat fruit. With a wing span over a yard long, fruit bats decorate trees in Africa and Asia, dangling like black socks. On islands of the Pacific and Indian oceans, megabats have cousins called flying foxes, including the *Sauve souris* of the Seychelles. Some researchers, including Jack Pettigrew of Australia, reckon that the megabats are a form of primate. They have what has been described as a strong working relationship between the eye and the brain.

Rayner finds the microbats of Britain use a very short FM call—30 to 50 kilohertz is average. "This is very good for finding structures, say, if you're looking for an insect on a leaf," Rayner says. "Sometimes these calls are modified by little harmonics, with a little tail on the end. They know how high they are off the water. It's an altitude reading."

In addition to orientation during flight, bats use echo location to detect a meal. At first, their signal becomes shorter and more frequent, and they can recognize an insect by its wingbeat frequency. The sound of a moth warming up for takeoff is attractive, since it marks a sitting target. A second signal, of longer duration, indicates airspeed and direction. A UFO intrigued a microbat in Maryland a few years ago, when a young Rick Adams was playing with a frisbee. The rotation attracted the bat, and the fatal collision led Adams to his current research. Jeremy Rayner recorded a trick of the horseshoes: "The bat is flying, the target is flying. The bat can't tell how much the movement is due to its own movement or to its prey. Because this is a constant frequency, it's very sensitive to

Doppler shifts. So they change the frequency to a stationary object, the echo comes back, and they compare this to the frequency of the moving insect." Rayner recorded two overlapping sonagrams that measured the calls of a wild horseshoe.

Bats use different signals in the wild than they do in a lab. "We were misled at first," Rayner says. "You get much more variation when you record in the wild." Not all bats use echo location. Vampire bats use infrared detection, a strategy also used by vipers. Some bats employ whispering calls of low intensity, so that their prey is unlikely to hear them. The Doppler shifts, with fake insect wing beats, can be simulated in the lab. "But so far no scientist can figure out how to read the echos. But bats can." So can moths, which figured out a way to jam the signals by altering their wingbeat. "It's a classic example of an evolutionary arms race; The insects learned to identify the bat's sonar and avoid them."

The wingbeats of insects occur in oscillations, produced by fibrillar muscles that store and convert elastic energy, operating like a spring. In the case of a fly with a three-speed gear shift, a little extension resembling a tooth can move into one of three valleys among peaks. When shifting gears, an audible click occurs. The gears are used to increase or decrease the wing beat, resulting in a variation of speeds not unlike the gaits of horses. Small insects create totally different aerodynamic wakes than larger ones. "With smaller insects, frequencies of wing movement get much higher relative to size, and that changes the whole airflow. Also, insects are very light; the wing loading and ratio of weight to an area is much lower in insects. Compared to an insect," Rayner declares, "a bird is more or less an elephant with wings."

"A bird had to be built with basically the same bones and muscles of an anthropoid," he continues, meaning primates like ourselves. "It's got the same loads almost as we've got. So a bird's development was limited, with minor changes, nothing dramatic. You see all these things that a mammal was designed for, except those wings."

Teeth were among the ballast discarded by birds, and even their sexual organs diminished; the female has only one ovary, and the sexual organs shrink until breeding season, when, for example, the organs of starlings weigh 150 times their weight during atrophy. The latticework of their wing bones is airy and light, and bones throughout the system trimmed in weight to such an extent that the skeleton of a frigate bird such as you see in the Galapagos weighs a mere 4 ounces. The frigate has a 7-foot wingspan. The bones of a pigeon account for less than 5 percent of its total body weight, while its breast muscles alone contribute half. Every bone and quill surrounds a hollow tube. There is no bone marrow, only air. Lungs connect to an auxiliary system of air sacs that push oxygen throughout the skeleton. Consequently, a bird with a blocked windpipe can breathe through its feet.

For a long time, pterodactyls were considered batty in form. But a look at good pterodactyl fossils, where the membrane left an impression in limestone, showed their wings were independent of the hind legs, and long and narrow, like those of soaring birds such as albatrosses. Other parts suggested the same flight behavior as soaring birds—minimum flap, maximum vision—the latter told by well-developed optic lobes in pterodactyl fossils. That the pterodactyl ate a marine diet like that of an albatross is fairly certain, since several fossils were found with fish remains in their gut.

In addition to the Q.N. fossils from Texas, pterodactyl fossils were found at the Solnhofen site in Germany and what is now Kazakhstan, once a part of the USSR. They were divided into two groups. The older ones had a long tail; the younger, more recent, ones had a very short tail, which Rayner thinks "played only a minor role in flight." Some of the fossil specimens were so well preserved that the fabric of the wing membrane made an impression on the limestone matrix of such detail that Rayner could see fibers, generally straight and continuous, but curving slightly as they approached the trailing edge.

Like the stress distribution of windloading from tree branches to the trunk, these fibers transmitted aerodynamic loads to the skeleton. The system was like that of modern birds and bats; feathers transmit the load for birds, and the load on a bat's wing transmits via the animal's bony fingers.

Sir Arthur Conan Doyle, the inventor of Sherlock Holmes, featured pterodactyls in his novel *The Lost World*, and fossil specimens were studied with intensity during an era when ornithopters were still on the drawing board. But the mechanics of their flight were misunderstood, because the very form of their wings was misunderstood. By studying the true profile, and the fiber pattern, Rayner reckons that pterodactyls had a wing structure that could "withstand being under unfavorable aerodynamic loads." In other words, he thinks they truly flew.

So does Kevin Padian, a paleontologist at Berkeley invited to consult on the Q.N. at a 1984 workshop organized by AeroVironment. "All pterosaurs were strong, active fliers, and only large size constrained this ability," he wrote. Padian thinks the larger ones relied on soaring. They probably evolved from flappers but maintained an ability to flap for takeoff, to gain altitude, and avoid danger. The low energy cost of soaring may have allowed them to grow "to a size at which flapping for an extended period of time is energetically impossible." Padian included a summary of other scientific papers with an unusual agreement on specifics. The pterodactyl was a "superb low-speed soaring animal that had difficulty flying in high winds and landing, but had a low sinking speed, an excellent lift/drag profile, a light wing loading, low turning radius, high maneuverability, and optimal performance at 7 to 10 meters per second. It was presumed to spend most of its time gliding at sea, trapping fish at the surface in its great beak. However," Padian adds, "its existence must have been marginal, because it was so large that it was only barely capable of level flight; how it managed to catch fish, recover from the weight of the prey, and overcome the sudden strain on the neck to rise above the water's surface was not clear."

The Q.N. replica flew and it didn't. Fifteen models were tested, including a replica positioned on top of a van, speeding along a California highway to test its flapping capability, looking like a bird of prey that had misjudged the size of its victim. The beak was used as a rudder, and the claws for banking and to reduce drag. Power was provided by 6 pounds of nickel cadmium batteries. The entire structure weighed 35 pounds.

In January of 1986, the AeroVironment team set out with their fake fur friend for El Mirage, a dry lake basin in the Mohave desert, a setting where, a few decades ago, you might have expected Chuck Yeager to appear on the horizon. The site near Edwards Air Force base was chosen to avoid Santa Ana winds. The first test flight was a simple glide. The second involved banking and turning. Both were successful. The Q.N. was launched again by a tow line, and this time it flapped its wings, banked, and headed west into the sunset, towards Hollywood.

The successful flapper made its debut in an IMAX film entitled *On the Wing*, and to publicize the film, a demonstration flight was scheduled at Andrews Air Force base, the home of Air Force One. Shortly after take-off, the Q.N. went into a nose dive; it then managed partial recovery, only to lose its head. Paul MacCready was not totally surprised. "The plate holding the head on had a crack in it, and apparently that was the weak spot." MacCready mounted the pterodactyl head on the wall behind his desk. "You have to try things out," he said later, "and accept failure as an opportunity to learn." That's the noble stance, but Mac-Cready confides that he felt uncomfortable about making that flight before the public "because there was no opportunity to check out all the variables beforehand." While the crack was the result, the original cause of the crash was a "spurious radio signal," that released the Q.N. "several seconds before its 'brain' could be activated and equilibrium established. After it spun," MacCready continues, "a parachute was released. There was obvious damage on landing, which we repaired in about two hours. Afterwards we joked that this almost-real creature had decided to make the supreme sacrifice so as to get its picture on the front page—and thereby popularize the film."

It was a structural failure, after all, a simple crack, not a theoretical failure, since the replica had flown over the Mohave long enough to be filmed. The word "disaster" was applied to the demonstration flight, and the *New York Times* ran a front page story "Pterodactyl Remains Extinct." "Flying Machine Fails" could have been the headline for the December 14, 1903, flight by Wilbur Wright, who ploughed into the sand shortly after takeoff. His brother succeeded three days later.

Pterodactyls have a flight history of 100 million years, compared to less than 100 years for humans. Kevin Padian recalls that interest in their aerodynamics began long before the Wright Brothers. "Early work particularly reflected the hope that pterosaurs would reveal possibilities for human flight, though as soon as workable aircraft were invented interest in pterosaurs quickly cooled. In the past decade it has been rekindled by the opposite hope, that modern advances in aviation might reveal how the pterosaurs flew." Advances such as strong, lightweight material helped the Q.N. fly, but what a leap to imitate an extinct bird, when some living birds still confound the experts. Jeremy Rayner thinks the "conspicuous lack of success" of ornithopters was a result of people misunderstanding the mechanics of airfoil action.

Wings are one of those multifunctional forms that we have used only in a linear manner, when their dynamics are far from static. MacCready considers modern aircraft to be "very static, dead shapes." The pterodactyl, on the other hand, had "many moving parts, in complex relationships," a point that German paleontologist Dolf Seilacher makes when regarding multifunctional forms. There is a maxim for machines: the fewer parts the better. This arose partly because parts broke; replacing them was inconvenient, costly, and frustrating, because manufacturers changed models and what you had was suddenly obsolete. The maxim can be maintained with parts more pliable, like Mattheck's screw, and, with his work on electric cars, Paul MacCready continues to demonstrate the novel rearrangement of parts that he honed with the Gossamer Condor. Redirecting the energy lost to braking could have been done fifty years ago. The same creativity led Frei Otto to repair bridges with limited materials, and such resourcefulness occurred on a journey across the Sahara, when a pair of my panty hose were substituted for a broken fan belt.

The novel rearrangement of available parts was a key theme in the work of D'Arcy Thompson.

4. The Legacy of D'Arcy Thompson

THUS far, no one has convicted Sir Peter Medawar of hyperbole for suggesting that D'Arcy Thompson was "an aristocrat of great learning whose intellectual endowments are not likely ever again to be combined in one man." These words were published in 1958, ten years after Thompson died, and Medawar's assessment still holds water because the strengths he listed also reflect Thompson's flaws. His aristocratic approach led him to dismiss some of his colleagues' work as "vulgar" and "uneducated." He boasted about "not running with the pack," and the pack noticed. His combined disciplines had obit writers and good friends struggling with whether he was a classical scholar who dabbled in geometry, or a pretty good biologist who could write.

"The mathematics available to D'Arcy Thompson could not prove what he wanted to prove," wrote James Gleick in *Chaos*. "Why then, was Albert Libchaber thinking about *On Growth and Form* when he began his fluid experiments?" The math to prove his ideas, multivariate analysis, was available in theory, but the speed to assemble the subtotals and try on every pair of shoes in the store, as it were, was not. It was a dilemma of destiny for Thompson. That math could explain the mysteries of life has propelled physicists ever since Newton refined a formula for gravity in 1667. Normally, it didn't propel biologists, to whom Thompson dared suggest that "numerical precision is the very soul of science." Winning numbers close the case, whether it might be Bucky

Fuller's patented formula for the geodesic dome—or when Newton acknowledged Robert Hooke's prior hunch regarding the law of gravity but said his own mathematical formulas proved the theory in greater detail.

Thompson counted everything in sight, steps to a church door, ripples in the sand, petals on a blossom. In Hong Kong and Tokyo, he surveyed local fish markets and conducted a census. How large were the fish, what kinds did people buy and how much did they pay, what were the populations of species in local waters, what size boats did fishermen use, how far out did they go, what was the case last year, and the year before. He did not confine his studies to partitions of space but had a grasp of the greater flux, dynamic systems that moved, forms of an oceanic and atmospheric scale, populations and the weather. He was a connoisseur of rare books and common weeds. When he went for a walk, he came back with the makings of a small exhibit in his pockets, and he established two museums rich in specimens. Details on the landscape weren't just collected, counted and described; they were folded into a community of principle.

Thompson's book might have been called *On Form*, but he added *Growth* to emphasize that life was not static, that forms responded and had to expand. The blueprints he saw were a diagram of forces, growth vectors, and geometric patterns for transformation as happens with the hexagons of the sand dollar. His influence on Dolf Seilacher is obvious; Thompson described the test of a sea urchin as "a membrane stiffened into rigidity," and wrote that "the rigidity of the shell is more apparent than real, for the entire structure is, in a sluggish way, plastic." He compared the shapes of sea urchins to drops of water sitting on a flat surface, their domes shaped by surface tension, their oblateness informed by gravity, which Seilacher repeated with balloons as models.

His words had an irrefutable influence on Chaos, which he called Interference. He saw order in fluctuation. The patterns that he discerned as diagrams of forces were not confined to organisms, but embraced action, including the dispersal of ink drops in a glass of water, where, unfurling like an eddy, they imitated the vortices that flow in a pigeon's wake. Thompson compared this to the movement of planes and ships, with drag diminishing as the drop itself tapered down.

He looked for regularities in the same way that led Mandelbrot to find fractals in irregular coastlines, and he made similar leaps, comparing the effects of a bubble bursting and the splash of a single drop. The formation of a crater ridge, demonstrated in the classic photographs by Harold Edgerton of a drop of milk splashing, represented an "instability," a key word for the theory of self-organization, which embraces a process that cannot reverse itself. The ink dispersed in water does not reassemble itself. The same is true of craters formed by meteorites making a gargantuan splash, and the eruption of a volcano. While Edgerton's photographs are legendary, and they were included in the second edition of *On Growth and Form*, published in 1942, Thompson noted that "rapidity" of photography was not required to capture the phenomenon. The coronal edges of craters could be demonstrated by drops of water splashing onto dry sand, or by the record a bullet left in soft metal.

No one else on the landscape drew up so many blueprints of nature, or made such an eloquent argument for their utility, partly because Thompson collected just about everything anyone else had done and pulled it together. For all the excellent literature on natural history, his book still provides a new way at looking at nature, with lucid explanations for economy and beauty. The patterns that emerged were basic, fundamental, of and from the roots of everyday life, like the radical laws that R.W. Emerson seized upon when he wrote, "Each creature is only a modification of the other; the likeness in them is more than the difference, and their radical law is one and the same." As a student of the sixties, my first impression of radical law was oxymoron. But one definition of "radical" gets close to a radish, the pungent root—"radical" means exactly that, going to the root, the foundation, the source of something. It is also the sign for delving roots. Radical laws link people with other forms, including the botanical, even the inorganic. A Greek column, a tree trunk, and a leg bone are all informed by gravity. Readers with no professional interest find a personal one. Thompson took the ancient longing for finding meanings in patterns and managed explanations that did not require logic to disappear in the Twilight Zone. You could compare the swirl of a fern pod to a galaxy, or a nautilus shell to your own inner ear. It was something you could wrap your mind around, and Thompson's simple black-and-white drawings were powerful.

When Thompson cast Cartesian grids over the outlines of fish, you could see how the whole form, from one species to another, changed as if by shearing. The impact was as direct as looking a gorilla in the eye,

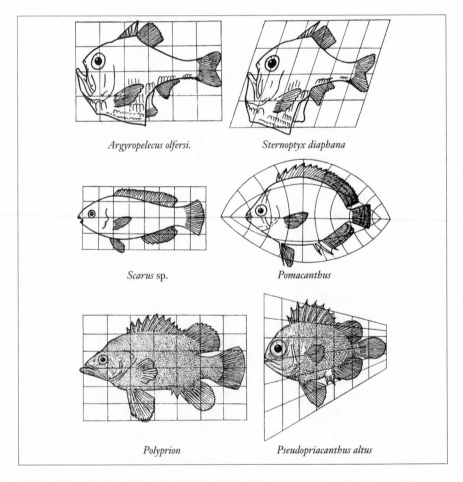

Argyropelecus olfersi.

Sternoptyx diaphana

Scarus sp.

Pomacanthus

Polyprion

Pseudopriacanthus altus

D'Arcy Thompson used the coordinate points of Cartesian nets to demonstrate changes in species. With only slight modifications to a single coordinate, one could obtain a whole new fish. The same procedure was used for evolutionary changes in a pelvis, the transition from a chimp to human head, and alligators throughout the ages.

with no need for an evolutionary tree. Had a paleontologist described the changes, they would have divided the fishes into parts. Thompson mapped changes as if the whole form were a globe and the changes were tectonic. Now the coordinates for the grid can be seen in 3-D, with new formulas to compute the changes in proportions. Thompson modeled in his head and with his hands, using threads and pins, tracing the patterns in pine cones, building models with cardboard.

He linked patterns to illustrate the influence of physical forces on forms. Surface tension dictates a Law of Minimal Areas, an economic use of space that has since been adapted in human inventions, from the hexagonal floor plan for warehouses, to the vectors for a geodesic dome, to the tensile wire meshes that Frei Otto used to shape the Munich Olympic stadia. If you dip a malleable wire into soapy water, you can contort the wire many ways, but the soapy membrane always maintains the shortest distance within the boundary because of surface tension. By playing with films, Otto's team at the Institute for the Study of Lightweight Structures came up with the German Pavilion at Montreal in 1967, a single membrane of undulating roofs supported by masts. A new possiblity space was found by dipping a string with a loop into the stretched membrane and pulling the string up, like raising a tent. Because soap film is inclined to reduce its area, it was a unique way to trim the use of material, and confirmed one of D'Arcy Thompson's themes: "The perfection of mathematical beauty is such that whatsoever is most beautiful and regular is also found to be most useful and excellent."

The "mathematical beauty" is apparent when you press soap bubbles between two glass plates. The result looks like a cross section of a honeycomb; all of the bubbles except those on the outer boundary assume a hexagonal pattern. The same angles occur in the cells of a plant stem; the result is a maximum number of cells in minimum space. They interlock, producing Y's with angles of attachments of 120 degrees.

The formula for soap films is known as Plateau's problem, after the Belgian physicist Joseph Antoine Plateau who wrote about it in 1873. Thompson put it alongside the Honeycomb problem, a pattern that has intrigued people for centuries and led to claims that bees were brilliant at math. The honeybee dazzled naturalists the same way a bumblebee defied aerodynamicists. Even Darwin gave honeybees credit for the "most wonderful of known instincts" and declared a honeycomb "absolutely perfect in economizing labor and wax." There were also notions that bees built circles but chewed the edges off and recycled the wax so they wouldn't waste material. Karl von Frisch, who won the Nobel prize in 1973 for his studies of how bees communicate, wrote: "It has often been maintained that the hexagonal shape was nothing spectacular since, under the influence of lateral pressure, the cells would acquire the shape with the smallest surface area. But the cells are made this way from the beginning. Right from the start, the cell walls meet at the correct angle of 120 degrees."

What was clear was that economy was the result. Thompson said these hexagonal cells were simply responding to physical forces; they begin round, like soap bubbles, but in their original semi-liquid stage, when the wax is warm and still has fluid properties, the 120° angles are formed as they attach in tension. The whole system is brought into equilibrium by the common angles. The result is known as the "close packing of cells," an arrangement not only found in plants, but also in the cornea of the human eye, dry lake beds, and polygons of tundra and ice.

From these patterns Thompson moved to flow, blood vessels, and the Law of Minimal Work, comparing the angles of branching, and the relative dimensions of the vessels. To this, he added the Principle of Similitude, which Galileo used to predict that a tree cannot grow above a certain height or it will fall from its own weight. It was a study of proportions, where surface area and volume influence behavior, locomotion, and even longevity. Small creatures live shorter lives; they move faster. Thompson assembled so many physical laws that it appeared as if behavior itself was a result of things like gravity, turbulence, and flow. The sand dollar suggests this is partly true. His fundamental aim was to challenge Darwinism, the notion that natural selection was simply the Creator on overtime.

The subtle but clear correction Thompson gave to Darwin's view of honeycombs was part of this radical theme. Forms, he said, are not simply the result of heredity or behavior, but are shaped by physical forces. It was a new way to analyze evolution. The legacy of D'Arcy Thompson is an evolving science of forms; his views of the greater flux influence not only Chaos theory but the new science of Complexity. His impact thrives in evolutionary biology, biomechanics and architecture. Frei Otto called his experimental architecture Form Finding. "The study of pattern is a new discipline, still forming," wrote Hugh Kenner in his 1973 book on Buckminster Fuller.

"The most powerful and passionate intellectual thrust today is a search for unifying concepts, for a metaphysical base that will again provide meaning for existence in a society that has all but lost it." So wrote curator George Nelson, in 1976, to introduce an exhibit at the Smithsonian's National Museum of Design in Washington, D.C. "We are discovering rather late in the day," Nelson continued, "that neither science nor technology is capable of doing this, and the dusty attics of religion and philosophy are being ransacked for old medicines and recipes, for any bits and pieces to patch the holes in a leaky social fabric." Hugh Kenner disagreed: "Principles are never shy, never shrink away. . . . They manifest themselves . . . and whenever we learn one . . . we have hold of an imperative to rethink our action. This is called putting principles to use; it is also called technology."

The theme for the 1976 Smithsonian exhibit that Nelson introduced was "MAN transFORMS," and among the contributors to the companion booklet was Bucky Fuller, who wrote on Synergetics (the behavior of the whole is unpredicted by its parts), a unifying concept akin to Self-Organization and Complexity. Bucky Fuller had zeal for the same radical laws that D'Arcy Thompson conveyed over half a century earlier. "What science discovers but fails to communicate," Fuller wrote, "is that the technology of the Universe, which we speak of comprehensively as Nature, operates only as a complex integral of exact mathematical laws." Fuller argued that science and technology are capable of utopia itself. His interest in affordable housing led to his earliest domes, prefab and portable. His geodesic dome, patented in 1954, repeated a pattern found in the organic, defying the myopic view of another contributor, who claimed, "Spaces are a human creation."

Designs in nature were pretty hot at the time; in 1968, James Watson published his best-seller *The Double Helix*, describing a natural form "too pretty not to be true." As Stephen Jay Gould has pointed out, when Watson and Francis Crick were trying to solve the problem of the structure of DNA, they didn't sketch ideas on paper but toyed with models of spheres and round sticks. Crick had picked up this trick from Linus Pauling, whose "success was his simple reliance on the laws of structural chemistry."

In 1994, Vaclav Havel, president of the Czech Republic, said that "The only real hope of people today is probably a renewal of our certainty that we are rooted in the Earth and, at the same time, the cosmos." The former playwright cited the holistic view of the Gaia hypothesis, and another theory that suggests humans are "not at all just an accidental anomaly" but mirrored in the universe. He echoed a lament of D'Arcy Thompson's: "We may know immeasurably more about the universe than our ancestors did, and yet it increasingly seems they knew something more essential about it than we do."

The essentials that are often cited include the regard for nature held by Native Americans, thunderously adopted by Thoreau: "We have no wealth but the wealth of nature. She shows us only fathoms, but she is a million miles deep." Havel said the same of classical modern science, which "described only the surface of things, a single dimension of reality."

Thompson used a slide rule, a graduated circle, but gave it the same shortcomings that Petroski assigned to computer calculations and Havel assigned to the availability of data, as if the art of pondering was left to anglers and things that diminished thinking time diminished thinking. "Our astronomers have their clocks, and their graduated circles, and their tables of logs and periodic functions, and many other things which we could not do without, and which show (or so we think) our superiority over men of ancient times," Thompson wrote in an essay, "Science and the Classics." He quotes a Dr. Fotheringham: "'We have every advantage that the Hare had over the Tortoise; but our equipment loses much of its value because (like the Hare) we haven't patience enough.'" He predicted the peril of innumeracy, a complaint that Bucky Fuller would repeat: Science has become so remote that we have lost our feel for navigation. The "ancients knew the stars far better than we do," Thompson wrote. "Our city streets shut out the sky, and new lamps blind us to the old. Our calendar is ready-made for us, and we ask not how."

Thompson asked how numbers were made. In the Pythagorean school, numbers *were* things. Pebbles were arranged geometrically, producing plane and solid figures. Number 1 was a single point, number 2 created the shortest distance between two points, 3 pebbles made a triangle, 4, a pyramid, and so on. A Greek geometer and precursor to Plato, Pythagoras's central message was that "things are numbers." This was extended to all forms, including the harmony of the heavens, numerology, and astrology.

Thompson used Pythagorean numbers as a tool. He was interested in the warping of geometry, not the philosophy for a perfect circle. The earth was a warped sphere, just as its elliptical path was hardly a perfect circle. Their warp could be explained by physical forces, and so could the shape of a drop of water. His point was that physical forces dictate the basic blueprint. The pebble he tossed is still making ripples.

Steve Vogel of Duke University wrote, "Scratch anyone in biomechanics and you'll find someone who read Thompson at an impressionable age." Stephen Jay Gould finds Thompson's ideas on organic form "curious in places, almost visionary in others, and always profound." In addition to expanding Thompson's vision in "The Spandrels of San Marco and the Panglossian Paradigm," authored with Richard Lewontin, Gould has drawn upon *On Growth and Form* for numerous essays that appear in *Natural History*. The evolutionary biologist from Harvard shares Thompson's regard for historical reach, multidisciplinary approach, and a fondness for footnotes. Both express admiration for Darwin, and both have added refinements to his theory of natural selection. (With Niles Eldredge, Gould proposed the theory of punctuated equilibrium, challenging Darwin's idea of gradual changes in evolution.) Thompson thought that mutations and heredity were not the only causes shaping nature, and to include physical forces he repeated Aristotle, whose idea of multiple causes is as useful in explaining human nature as it is nature. This was Thompson's genius, to dust off an old idea, and give it new life.

D'Arcy Thompson is often described as ahead of his time, yet the Scottish biologist born in Edinburgh in 1860 was perfectly suited for a number of previous eras, and spoke languages considered dead. Some of his earliest conversations with his father were in Greek and Latin, which led to a high regard for context: "A fact discovered yesterday is balanced by the history of two thousand years," he once wrote. His survey of patterns stretches back to Aristotle and Archimedes, an accumulation of ideas assembled like the layers of the earth geologists call horizons. His horizons went far beyond the city of Edinburgh to include oceanography and the marine mammals of the Bering Sea, the locomotion of whales and whirligigs, the geology of the Carboniferous, the cosmos of numbers, the structure of dinosaurs, and the trusses of bridges.

He was aware that numbers, tortured enough, will confess to anything, and in his quest for precise roots, dug into some ancient calculations for the movement of the sun and the moon and found them "extraordinarily, almost incredibly accurate," uncovering errors of only half a second. He embraced the broad, scholarly view of the Victorian naturalist, so broad that it extended long after Queen Victoria's death in 1901. Even as Darwin regarded the beaks of finches, he was not an ornithologist, nor did his work with orchids fix him as a botanist. Disciplines weren't so delineated, and this appealed to Thompson. Like Darwin, he hesitated to publish his ideas. The two men were contemporary on a cusp; Darwin died in 1882, when Thompson was a student at Cambridge. Both were prescient, and it says something for vision that both focused on the past. The lag we perceive may be current; great strides occur in technology and data, with fewer steps in thinking.

"In my view it is utterly wrong to regard D'Arcy Thompson as a biologist with a good knowledge of classics and mathematics, or a classical scholar with a good knowledge of mathematics and biology," wrote Thompson's friend, zoologist Clifford Dobell. Sir John Myres, in an obituary for *Nature*, had described Thompson as a zoologist "who kept up his classics," which Dobell found "very wide of the mark." To distill the matter, Dobell suggested that his disciplines were chemically combined. "And he was a born writer." Yet his daughter, Ruth Thompson, describes him as a rewriter, a revisionist who spent the wee hours in search of the right metaphor and phrase. He fiddled with words forever. His papers were late, his manuscripts were late, and it took him nearly twenty years to revise the second edition of his book. In the 1942 edition, he failed to correct some wrong ideas, and he omitted research that pertained.

"The twentieth century's revolution in biology, well under way in his lifetime, passed him by utterly," wrote James Gleick. "He ignored chemistry, misunderstood the cell, and could not have predicted the explosive development of genetics. . . . No modern biologist has to read D'Arcy Thompson." Gleick continues: "Yet somehow the greatest biologists find themselves drawn to his book. This classicist, polyglot, mathematician, zoologist tried to see life whole, just as biology was turning so productively towards methods that reduced organisms to their constituent functioning parts. Reductionism triumphed." D'Arcy Thompson resisted this deconstructionist approach (as Gleick remarked, "D'Arcy Thompson saw this coming"), and it is hardly coincidental that a return to an interdisciplinary approach coincided with a return to a holistic view of forms.

When writing about the close packing of cells, for example, Thompson cited Sach's rule, based on the botanist's 1887 lecture at Oxford: "The behavior of the cells in the growing point is determined not by any specific characters or properties of their own, but by their position and the forces to which they are subject in the system of which they are a part. This was a prescient utterance, and is abundantly confirmed." It had the ring of Synergetics, Self-Organization, and even Complexity, where such patterns have become known as cellular automata in the lexicon of computer programs. From a 1994 article on Complexity: "Cellular automata work on a lattice, each "cell" of the lattice changing contingent on what it knows about itself and what it learns about its neighbors. They have been utilized to simulate processes like crystal growth or the intricate patterns seen on mollusk shells." Thompson considered the hexagonal patterns of snow flake crystals as "visible proof of the space lattice on which their structure is framed."

While modern biologists don't have to read Thompson, G. Evelyn Hutchinson of Yale suggested that a cheap reprint of the chapter "On Magnitude" ought to be used for classes in biology, mechanics, and English, since "it deals with fundamental and elementary matters which have been known, and have been unappreciated, for a very long time."

You could read some passages to a child. Thompson explains the similarities between "a little porpoise and a great whale," and why an oak tree does not grow as tall as a pine. (The oak carries a heavier load.) On velocity: "We comprehend the reason why one may always tell which way the wind blows by watching the direction in which a bird starts to fly." Just when you think you're going to be knee deep in stress and strain, he draws a kitchen analogy of kneading dough to explain "stimulation by pressure." Don't be put off by the first page, a dithyramb to numbers, and biblical in the sense that one geometer appears to have begat another.

Physicists get hooked in the introduction: "The waves of the sea, the little ripples on the shore, the sweeping curve of the sandy bay between the headlands, and the outline of the hills, the shape of the clouds, all these are so many riddles of form, so many problems of morphology, and all of them the physicist can more or less easily read and adequately solve. . . . Nor is it otherwise with the material forms of living things. Cell and tissue, shell and bone, leaf and flower, are so many portions of matter, and it is in obedience to the laws of physics that their particles have been moved, moulded and conformed."

"It is easy to assess D'Arcy Thompson's influence," Gould wrote, "for that is illustrated by the continuing use of his work in the technical research of distinguished scientists, and by the location of his main ideas at the core of an emerging science of form."

A small measure of this influence is a booklet published by SFB 34, a citation index for references to Thompson's work from 1970 to 1989. A citation means that one of his ideas was used, tested, even abused or refuted, and the author(s) gave him credit. Most are scientific papers.

References in the popular press (magazines, books, newspapers) were not included, and the source was limited to *On Growth and Form*, eschewing any references to his writings on Aristotle, his comprehensive glossaries of Greek birds, fishes and fauna, his essays on "Science and the Classics," and so on. The citations total 610.

There are reports on shells, and form in general, yet the applications are, as Ekkehard Ramm once said of the SFB conferences, "extremely widespanned and exotic." There are papers on Roman arches, human joints, and disease; simulator design and instructional features for air-to-ground attack; elemental carbon cages (now known as Bucky balls); fractals in physiology and medicine; facial effects of fetal alcohol exposure; and cosmic strings as random walks. These 610 citations do not include those after 1989 (the annual SFB and SEB conferences, for example), or before 1970, such as *Essays on Growth and Form*, edited by Sir Peter Medawar and Le Gros Clark, published in 1945, to honor Thompson's sixty years as a professor.

Some of his ideas were wrong, as happens when you have the "audacity for imagination." He was wrong about the cause behind the shape of a hen's egg, and as Gleick notes, he misunderstood the cell. "The biologist in him, strangely enough, was the weakest member of the team," Medawar wrote. The famous leg bone with its diagram of forces (Cullmann's crane) certainly responds to gravity, but the fact that we no longer walk on all fours is a result of natural selection. Presumably gravity hasn't changed in the 4 million years of human evolution. In a sense, Thompson overstated his case, diminishing the role of heredity to make his point, the same way that Gould and Eldredge, in their first paper on punctuated equilibrium, virtually denied gradualism to impress readers that changes in evolution can be abrupt.

Critics nailed Thompson for not following up his ideas with experimental proof or using statistics in the book. His approach was comprehensive in terms of literature, but anecdotal in terms of science. "D'Arcy Thompson believed that more data had already been accumulated than any man had yet comprehended," defended Dobell, "so he did not apply himself to the collection of still more data, or to experimental work in the laboratory, but devoted his immense energies to the elucidation, analysis, and synthesis of facts already ascertained."

Written in what has been described as the bel canto style, his sentences are lyrical compositions, just shy of Faulknerian length. "To spin words and make pretty sentences is my one talent," he wrote, "and I must make the best of it." Late at night in his study, he would pace back and forth, repeating a passage aloud, keeping time with the wave of his hand as he spoke. He was composing a paragraph, tuning a meaning. When the sound suited him, he returned to his typewriter.

D'Arcy Thompson loved the opera, waltzed until the orchestra called it quits, and lectured with such impact that students remembered his words half a century later. These details of his personal life were recounted by his daughter, Ruth D'Arcy Thompson, in a biography. *D'Arcy Wentworth Thompson, The Scholar-Naturalist*, was published by Oxford University Press in 1958, and includes the comments by Sir Peter Medawar in a postscript. I have combined quotes from this book with Thompson's lectures, essays, and papers to reconstruct aspects of his personal and professional life.

The rapport of inquisition that Thompson cultivated with his daughters suggests that he had the wisdom to think like a child. (A family photo in Ruth Thompson's book features another daughter, Barbara, with the caption: "D'Arcy and Barbara looking at beetles on the Perthshire moors.") In addition to his scholarly works, he wrote two books of riddles (*Rhymes With Reason*, and *Rhymes Without Reason*) that were blockbusters in nurseries throughout Great Britain. He was over six feet tall, wore a beard, wire rims, and tweeds until they were smooth. His eyes were described as a penetrating blue, and his hair was red until he began to publish what he called his "heresies," around the age of 55.

As a student he spent his Saturdays "botanising" around the countryside with the Eureka Club, whose members included J.S. "Johnnie" Haldane, who remained D'Arcy's friend for nearly seventy years. Haldane studied at Oxford and became a physiologist, to work alongside Thompson at Saint Andrews College in Dundee, Scotland, but he is best remembered for producing a son, J.B.S. Haldane, the geneticist who contributed to the Modern Synthesis, a multidisciplinary approach to evolution.

When D'Arcy Thompson "botanised," he also birded, geologized, mollusked, and captured butterflies with a homemade net. Ferns went into a handkerchief, sea shells into his pockets, seed pods into his umbrella. His interest in anatomy began at the age of 9, when he helped his grandfather, Joseph Gamgee, a veterinary surgeon. Gamgee studied bone structure among certain breeds of horses at certain gaits, the changes in stress during a gallop, and fatigue in the lame. D'Arcy developed a regard for the tension and compression forces in tendon and bone. It would be the subject of his first thesis in college, and the influence of loads on a bone was pivotal in shaping his ideas about the diagram of forces. His thesis at Cambridge was "On the Nature and Action of Certain Ligaments," which Medawar cites as "evidence that he was interested in bones for how they worked rather than for what they might have to say about their owner's evolutionary credentials." Thompson complained that skeletons in museum exhibits were misleading because they featured only bones. The true mechanics of the structure required tension lines as well as compression lines. He admired the ancient practice of anatomists who left ligaments attached, and made the point that in the organic, tension and compression work together.

At Edinburgh Academy, Thompson encountered an instructor who conveyed the rule of great teaching, which Thompson would transpose to the rule of great writing, and the study of patterns, "that nothing is interesting by itself, but that things become 'interesting' as soon as we have stories to tell about them and begin to weave one thing into another." His intrigue with patterns in shells began with a friend of his father's, a mathematician keen on the geometry of the Greeks. "From the Naturalist's side I should say that he knew little, or even nothing about [mollusks]. . . . But he seemed to take an intense pleasure in their beauty; . . . he found something at least in touch with his intelligence, something which a mathematician could enjoy and appreciate and understand."

In the early 1600s Descartes had sorted out the logarithmic spiral, which differs from an even coil such as you find in a cinnamon roll, or a nautical coil, of uniform thickness. In a log spiral the radius increases by the same factor in every turn. For a nautilus, these increases are graduations in real estate.

The shell of a nautilus, like the nautilus itself, grows in size but doesn't change its shape. The young shell is a scale model of the mature one. "In the growth of a shell," D'Arcy wrote, "we can conceive no simpler law than this, namely that it shall widen and lengthen in the same unvarying proportions: and this simplest of laws is that which Nature tends to follow." The geometric result is equal angles from the radius, measured from the starting point of the shell. Every angle is proportional to the logarithms of the successive radii. Logarithmic spirals are also called equiangular spirals. Descartes laid down the formula; James Bernoulli named it the logarithmic spiral, and it applies to galaxies, the horns of a ram, snails, and, as Thompson discovered in 1888, microscopic foraminifera.

These sea creatures collect carbonate of lime from salt water, and concentrate it to calcite. Like the nautilus, they start with a single chamber, and add more. The dividing walls between these chambers have holes in them, which explains their name: the Latin *foramen* means hole, and *ferre*, to carry or bear. Foraminifera are so plentiful and tiny that on some beaches, 50,000 have been found in a single gram of sand.

Thompson distinguished their log spiral from that of an elephant's trunk or a chameleon's tail, "a transitory configuration" of muscular forces. The spiral in a fern pod, however, is the result of growth, along with incremental growth that occurs in tusks (as in elephants, wart hogs, and walruses), and in claws (in cats, birds, bears, and pottos). Even though tusk and fingernail cells are dead, they extend by secretions or deposits from living cells.

While their form is symmetrical, their growth is asymmetrical. Nautiluses and foraminifera grow from only one end. The same is true of horns, adding rings from the base, creating two generating curves that mirror each other in symmetry. This increase by *terminal* growth, with no change to the form of the whole figure, is "characteristic of the equiangular spiral, and of no other mathematical curve," according to Thompson.

This portrait of D'Arcy Thompson with a nautilus shell was taken by Björn Soldan in 1946, and three years later, featured in Thompson's obituary.

D'Arcy Thompson's ideas about the diagram of forces were not expressed in print until 1915, when he published "Morphology and Mathematics," a trial balloon for *On Growth and Form*. Yet even as he entered Edinburgh University as a medical student, an instructor in physics influenced him with accounts of the patterns of rainbows and the aurora. In addition to physics, paleontology intrigued him, and he lectured to the Edinburgh Naturalists Field Club on the Carboniferous. By 18 he was financially independent, teaching Greek and publishing articles. Medical science lost him to Trinity College, Cambridge, where one of his mentors was Frank Balfour, head of the Museum of Zoology, and only a few years older than his students. "One great branch of biology actually grew up in his hands into a science—the science of embryology," Thompson said of Balfour. "He gathered up all the scattered knowledge that was hidden in books and floating in men's brains, and he, for the first time, wove it together, and entwined with it the thread of his own originality and genius." This approach to knowledge became Thompson's own blueprint for his teaching and writing.

During his first year at Cambridge, the publisher Macmillan commissioned him to translate *The Fertilization of Flowers*, written by German geneticist Hermann Müller. (Macmillan had initially suggested a second German work, on hybridization, which led Thompson to venture later that perhaps he could have anticipated Mendel's insights by a couple of decades had he translated that work, too.) Charles Darwin was asked to write the introduction to *The Fertilization of Flowers*, and he welcomed the translation since he was not fluent in German. Based on his own backyard experiments with orchids, Darwin had written "On the Various Contrivances by which British and Foreign Orchids Are Fertilized by Insects." It underscored his own regard for form and function, and was published in 1862, when D'Arcy was 2.

Organic form and function, from orchids to the beaks of finches to variations among giant tortoises, was essential to Darwin's theory of natural selection. Ironically, his followers deemphasized such design, partly for reasons of semantics, for design implied (as it still does today) no malleability. Darwinism is often mistaken for Darwin. What Charles

Darwin actually thought or wrote was often misconstrued (as in the doctrine of social Darwinism and notions of survival of the fittest). Phylogeny and taxonomy dominated Darwinism, and forms were studied for clues to classification. In the quest to classify, organisms were disassembled and described, part by part, bone by bone. Reductionism triumphed.

D'Arcy Thompson was an admirer, yet even as the two men corresponded when he was at Cambridge, he was developing ideas that would challenge a basic tenet of Darwinsim. The introduction to *The Fertilization of Flowers* was one of the last things Darwin wrote before he died. The two never met.

In 1885, a year after his graduation from Cambridge, Thompson said in a lecture in Scotland that "it is a mistake to suppose that evolution has yet settled down into a consolidated scheme or that those who work by its light are pledged to the stereotyped clauses of an evolutionary creed." Nearly a decade later, he lectured on "Some Difficulties of Darwinism" at Oxford.

In hindsight, Thompson described Balfour's approach as too "engrossed in the problems of phylogeny." Forms were studied for evolutionary links, to draw up the branches of lineage. How these forms functioned was virtually ignored, but D'Arcy Thompson had maintained his own private study, debating, via letters, with his Grandfather Gamgee about how a bone is built. When he first submitted his ideas to Michael Foster at Cambridge, the response was, "If the form is constant in a group—it does not matter how the form is brought about."

At Cambridge, he began a second book, a bibliography of protozoa, sponges, coelenterates, and worms (covering some of the earliest blueprints in nature), and then proofed the second edition of a dictionary of the English language. D'Arcy began to collaborate with his father translating Aristotle's *Historia Animalium*, a project that was shelved until 1910, presumably so that D'Arcy might plumb books other than the ones he was writing himself.

His Uncle Gamgee warned D'Arcy about doing too many things at once: "As the Italians have it, *chi troppo abbraccia nulla stringe* (he who attempts too much achieves nothing). At the same time, his grandfather advocated a multidisciplinary approach. "You know that I conceive no course of teaching of the two halves of one subject (I mean Anatomy and Physiology) that will not suffer as whole things all do when split into halves. The new fangled idea of subjects being so great that only parts must be undertaken by one man is a consummate absurdity." Thompson took on everything, wavering and divided in his purpose, as if he resisted being classified into a group himself.

In addition to his publications, his tutoring in Greek and biology, and his post as junior demonstrator in the Cambridge Science School (for £30 a term), D'Arcy helped establish Toynbee Hall, a settlement house in the slums of London, ran a Boys Workshop, and passed his exams. His attempts at too much caught up with him when his application for a fellowship was rejected. He settled for a post at a small college in the port of Dundee, where the oldest professor was 31. At the age of 24, Thompson chaired the biology department, teaching anatomy and zoology, vertebrate morphology, botany, and comparative embryology; lab work went on until ten in the evening. He established a museum, working after hours to prepare 600 specimens, cajoling local sea captains to deliver specimens from Siberia. Then he established a medical school. His old classmate J.S. Haldane joined the staff, investigating what is known today as the phenomenon of "sick" buildings; Haldane examined respiratory problems in slums, submarines, and coal mines. Thompson's many interests diverted him from his research into the geometry of patterns, although in 1888 he set out for Paris to "pay a visit to Tramond, the skeleton man, and on then to Collin, the great instrument maker." A year later, he was elected a Foreign Member of the Moscow Academy. ("On the strength of this I think of ordering a new suit of clothes!") Whether it was the new suit, Russian inspiration, or cumulative guilt, he wrote that he had taken to mathematics, and discovered some unsuspected wonders in foraminifera.

One of his students recalled, "I shall never forget his description" of the angles of sponge structure "because for the first time I became aware that mathematics may be applied to give precision to biological observations." Professor Michael Foster, his advisor at Cambridge, was less enthusiastic: "To be candid I think this is too polemical. . . . I suppose everyone must admit that there are 'laws of growth'. . . But if so the argument should be cumulative, pointing out in case after case, that something behind natural selection is at work."

This was exactly the received wisdom that Thompson wanted to undo. It was as if natural selection had shaped the clouds and the moon. Perhaps as a reaction to these discouraging words, Thompson sought the biggest space he could find, and went to sea on a sabbatical. The rhythms of the ocean would influence him. The British prime minister, Lord Salisbury, appointed Thompson to the Bering Sea Commission because of his expertise on marine animals. The issue was the slaughter of fur seals near their breeding grounds on the Pribilof Islands, and a trade agreement between England, France, and the U.S. Thompson sailed via Bangkok and Hong Kong, had sushi in Tokyo, and traveled by dogsled on the Bering islands. That cleared the cobwebs, along with a book he read on his journey. *The Will to Believe* contained an essay on Determinism, which inspired Thompson to entertain an opposing theory of Interference. "We see indeed the sort of evolution of chance, and ever increasing complexity of accident and possibilities. One wave started at the beginning of eternity breaks into component waves, and at once the theory of interference starts to operate." The letter containing this thought was never mailed; the words suggest the essence of Chaos theory.

This notion led him to compare the patterns in a rolling surf to minor splashes. He got into flow, and its interference—viscosity and turbulence. This dynamic overview, the complexity of accident and possibilities, led physicist Albert Libchaber to an experiment that demonstrated "the infinite cascade."

"The leap from maps to fluid flow seemed so great that even those most responsible sometimes felt it was like a dream," wrote Gleick in *Chaos*. A map expresses flows in the meander of rivers, and fractals in the coastline. Thompson was accustomed to plucking patterns from a map, which he described in his essay on the invention of the blackboard in "Science and the Classics." He recalled his favorite book as a schoolboy, *Classical Geography*, by James Pillars, who "found the ordinary maps redundant with detail; he wanted to show 'the great natural features and palpable realities' and only such few towns, rivers and tributary streams as formed the scene of his story." To do this, Pillars built the first blackboard from beechwood, and mixed ground chalk with a teaspoon of oatmeal. Then he drew only the parts of the map he wanted to emphasize, a practice that D'Arcy Thompson employed. "When I tell my students about the far-flung migration of eels, or try to explain the order which underlies the scattered distribution of coral reefs in the oceans of the world, I do as Pillars taught me. I draw my chart of the 'palpable realities.'"

Other realities confronted him. When he questioned the Darwinian approach in a lecture, an old don took him aside: "D'Arcy, you may think these things, but you must not say them. It is not the time, and what is more, it is not the way to get on." His colleagues didn't support him for a post at Oxford, and other applications to universities were rejected. Around the turn of the century, he was appointed the scientific advisor for the Fishery Board of Scotland. The Board not only dealt with the trawling trade in local waters, but relied on deep sea exploration of the British research vessel the HMS *Challenger*. He investigated the currents and tides, the stratification of waters of different temperatures and salinity, seasonal fluctuations, and migrations. He used statistics to record the size and number of fish, and a record of deep sea temperatures over a long period of time. He superimposed isobars to illustrate the latter and devised other curves for the patterns of the tides. He considered that a fish scale tells its owner's age. From the bounty of the *Challenger*, he examined the drawings of Ernest Haeckel, of hundreds of skeletons of Radiolaria, "geometrical forms of peculiar elegance and mathematical beauty." He found latticeworks of hexagons and tetrahedrons, including one with an inner tetrahedron, like a bubble within a bubble. He found icosahedrons, with 20 triangular faces. He found geodesic structures.

◆ ◆ ◆

IN the early 1900s, most of Thompson's papers were rejected. "Plato's Theory of the Planets" was a hit in the Greenwich *Observatory*—the only hit between 1900 and 1910. His position "on the fringes of legitimate science" noted by James Gleick was an observation also made by his daughter, Ruth Thompson: "He hovered, as it were, on the fringes of both the scientific and the classical worlds, making, apparently, no deep impression on either."

Ruth Thompson noticed a change after the age of 50, when D'Arcy Thompson was able to let go of his father's bent against "worldly success" and recognition after his father died. His first recognition was ironic. In 1911, Thompson received a doctorate from Cambridge, not for science, but for literature. His translation of Aristotle's *Historia Animalium*, begun thirty years earlier, had finally been published. That same year, at a meeting of the British Association in Portsmouth, he attacked the dogma of Vitalism.

It was dogma meeting dogma, since all changes of form are not due to the action of physical forces. Vitalism is also known as teleology (from the Greek *tele*, meaning end, far away, terminal), the focus on the final form.

"To seek not for ends but for antecedents is the way of the physicist," D'Arcy wrote, "who finds 'causes' in what he has learned to recognize as fundamental properties . . . or unchanging laws, of matter and of energy. In Aristotle's parable, the house is there that men may live in it; but it is also there because the builders have laid one stone upon another."

Steve Gould breaks down the causes that built Aristotle's house: "the stones that compose it (material cause), the mason who laid them (efficient cause), the blueprint that he followed (formal cause), and the purpose for which the house was built (final cause)."

The stones that compose a sand dollar include hexagonal plates of calcium, shaped by heredity and the record of physical forces. The formal blueprint was a full dome, altered by adaptation; the house, not yet final, because it can change again.

Thompson's diagram of forces was a concept more palatable to engineers than biologists. A former engineer named Herbert Spencer had set the stage. Thompson mentions Spencer's work, *Principles of Biology*, in the chapter "On Magnitude": "It follows at once that, if we build two bridges geometrically similar, the larger is the weaker of the two, and is so in the ratio of their linear dimensions. It was elementary engineering experience that led Herbert Spencer to apply the principle of similitude to biology." Thompson continued the theme in his chapter "Form and Mechanical Efficiency," comparing the American bison to a "quadru-pedal bridge."

In 1913, Thompson gave the Herbert Spencer Lecture at Oxford, in honor of the philosopher, who died in 1903. He mentioned the "doctrine of the correlation of physical forces" that Spencer had featured in his *Synthetic Philosophy*, and traced the idea to Goethe, who wrote in 1795: "The more imperfect a being is, the more do its individual parts resemble each other. . . . The more perfect the being is, the more dissimilar are its parts . . . and subordination of parts is the mark of high grade organiza-tion.'" Thompson then described the work of his favorite biologist: "Though Aristotle follows the comparative method, and ends by tracing in the lower forms the phenomenon incipient in the higher, he does not adopt the method so familiar to us all, and on which Spencer insisted, of first dealing with the lowest, and of studying in successive chronological order the succession of higher forms. This historical method . . . the method to which we so insistently cling, is not the only one."

D'Arcy Thompson knew his view of life was an affront, and when he sent the manuscript of *On Growth and Form* to a friend at Cambridge University Press, he wrote, "I have tried to make it as little contentious as possible. That is to say where it undoubtedly runs counter to conven-tional Darwinism, I do not rub this in, but leave the reader to draw the obvious moral for himself." Meanwhile, his paper on "Morphology and Mathematics" was praised as "profoundly interesting," and the Carte-sian grids remain the most original concept in his book. A reviewer in *Nature* wrote of *On Growth and Form*, "It is like one of Darwin's books, well-considered, patiently wrought-out, learned and cautious—a disclo-sure of the scientific spirit."

Thompson lectured throughout Europe, in Russia, in Boston, Phila-delphia, Princeton, and Capetown, where his subject was "Anatomy from an Engineer's Point of View." In Delhi, he walked to the podium with a live hen under his arm. He was the host of a radio program on nature, and his collected essays were published in *Science and the Classics;* "The fact that I have loved them both has colored my life, and enlarged my curiosity and multiplied my inlets to happiness." In 1937 Thompson was knighted, and at the age of 85 he received Oxford's highest degree. Sir D'Arcy Thompson died at the age of 88, on the longest day of 1948.

As luck would have it, the first edition of *On Growth and Form* was published during World War I and the second edition during World War II. He expanded a 300-page book to 1,100 pages, which does not imply an improvement. The 1961 edition was edited by John Tyler Bonner of Princeton, whose decision to cut the chapter "On Leaf Arrangement, or Phyllotaxis," was the most difficult. "The main reason" for the deletion, Bonner wrote, "was that D'Arcy Thompson had really contributed no new information to this old subject, although his chapter is an excellent summary of the old views and the games with numbers."

The games with numbers have resumed. As Bertrand Russell put it, "Perhaps the oddest thing about Modern Science is its return to Pythagoreanism."

5. Divine Math

*Politics, of course, is a matter of decimals, logarithms, and long
division, but Mrs. Thatcher, by making parts of it appear simple,
not only infuriated those who knew it to be more complicated but
also cemented her support among the two-plus-two brigade.*
— JULIAN BARNES, "The Maggie Years,"
The New Yorker

I N 1980, in the Chinese city of Souzhou, I bought silk fabric by the
yard in a bustling store where sales receipts went flying to the check
out counter from points nearby and extreme. The receipts were attached
to cords, and reached a certain velocity at the level of my nose, qualify-
ing me for a near miss. I was aware of these busy vectors, but listened to
the sound of capitalism in China. Wooden counters provided percus-
sion. Each clerk used an abacus, sliding oval counters on wires in a frame
with blurred fingers. The lowest counter on the wire was one, the next
was ten, the third, a hundred, and so on; the upper level of this calculating frame denotes multiples below. The instruments were not strictly
musical, but when there was sudden pause, I half expected the clerks to
burst into song.

Liber Abacus, published in 1202, introduced Arabic numbers as a handy alternative to Roman numerals. The book's author was a 27-year-old named Leonardo Pisano. Leonardo Pisano had been educated by Muslims in the North African city of Bugia, on the Barbary Coast, but had returned to his native Pisa, thus his last name Pisano. That he was the son of Bonaccio led to his other name of Fibonacci, which unfortunately translates to Son of Simpleton. His nickname was scarcely alternative: *Bigollone*, the Blockhead. Such regard may have derived from his penchant for grafitti; he scribbled calculations on city walls. To people unaccustomed to Arabic numerals, his exercises must have appeared as strange and disjointed art, and even if recognized for what they were, most people recoil at a formula of any kind. Nobel prizes are not awarded in mathematics.

Fibonacci's book on the abacus included a small problem that gained in magnitude. It had to do with rabbits multiplying.

Take two rabbits of the opposite sex. Rabbits bear young two months after their own birth. Assume a pair of rabbits produces another pair every month over a year. By the end of the year, it was said, the original rabbits and their descendants would number 233 pairs. Fibonacci listed the total pairs of rabbits at the end of each month, and came up with this sequence: 1, 2, 3, 5, 8, 13, 21, 34, 55, 89, 144, 233.

The sequence began to pop up elsewhere. The petals of daisies, for example, usually have 21, 34, 55, or 89 petals, which is why it's better to begin with "loves me" rather than "loves me not." Pine needles often grow in clusters of 2, 3, or 5. The sequence in pine cone scales is 8 and 13, which also occurs in pineapples. It takes some doing to notice these things. Counting the 13 branches of a pussy willow requires you to revolve the stem, or yourself, five times.

Small sunflowers have spirals of 21 and 34 seeds, larger ones, 34 and 55. One spiral goes clockwise, the other counterclockwise. The seeds start small and new in the center, and grow out in a circle, larger and old. Many people see these as log spirals, yet unlike the nautilus shell or kudu horns, the growth is not connected in one form but has many increments. The spiral "advances in the direction of its own focus," wrote

The growth of seeds within a sunflower create logarithmic spirals, as do the scales of a pine cone. The log spiral is the only curve in math that enlarges without changing its shape.

D'Arcy Thompson, who did not think the botanical neat enough to measure its angles with precision, and included a picture of a cauliflower to make his point. Yet sunflower seeds are neater, and their spirals copied in straw hats with a brim and in the round dream catchers produced by the Lakota people of the American West, a lattice net of log spirals hung over the bed and thought to filter out nightmares.

"I, for myself see no subtle mystery in the matter," Thompson wrote, "other than what lies in the steady production of similar growing parts, similarly situated, at similar successive intervals of time." He saw the face of the sunflower as a shortened stem, a variation of a common generating pattern. This view was buoyed by a work of Goethe.

In 1790 Goethe wrote a paper about the development of plants, "Metamorphose der Pflanzen." He suggested that parts of "the vegetating and flowering plant, though seemingly dissimilar, all originate from a single organ, namely, the leaf." Goethe's theory has been widely misunderstood, because what he meant was not a current leaf available for plucking, but an archetype that might serve as a basic model. Nor did he include the roots or the stem. What he envisioned was an unfurling, with underlying principles, and that part of his theory suited Thompson. Seeds are closely packed for economy, branches, equidistant for balance and distribution. The spiral arrangement of leaves and branches enhances photosynthesis; Thompson wrote, "no leaf should be superimposed above another" to create shade. Others saw mystery in the matter.

When Fibonacci took a good look at the numbers 1, 2, 3, 5, 8, 13, 21, 34, 55, and so on, he realized that each number was the sum of the two preceding. When others took a look at the numbers, they found that if you divide a number by the next highest, it is always 0.618033... times as large as the number that follows. And 0.618033... is considered a "magic number." It expresses the Golden Mean.

To produce a Golden Mean, take a square and divide it down the middle, then make an addition to the right with an arc. The arc to determine the new lower right end is drawn by putting the point of a compass at the bottom of the dividing line of the square. The addition is slightly over half the original square, or 0.618033..., and is called the "golden proportion."

The Golden Mean is also called the Divine Proportion. The navel, for example, often marks a ratio of 0.618 of an adult's total height. In an infant, the navel is in the center, or divides the body in half. The golden proportion begins at about the age of 13.

The Greek letter phi (φ) is used to denote the golden ratio, after the Greek sculptor Phidias who emphasized golden proportions said to occur in human anatomy. Phidias made neat and divine the width of the throat to that of the head, the width of the ankle to the calf, the width of the wrist to the forearm, the narrow part of the thigh to the greater part, and so on. The terms Golden, Fibonacci, Divine, and Ideal became interchangeable, like symmetry and harmony were centuries ago. Now we think of symmetry as equal parts in balance, but in ancient usage symmetry implied pleasant spatial arrangements.

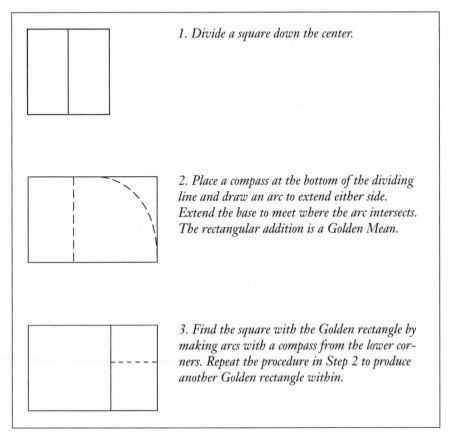

1. Divide a square down the center.

2. Place a compass at the bottom of the dividing line and draw an arc to extend either side. Extend the base to meet where the arc intersects. The rectangular addition is a Golden Mean.

3. Find the square with the Golden rectangle by making arcs with a compass from the lower corners. Repeat the procedure in Step 2 to produce another Golden rectangle within.

How to Draw a Golden Mean

Builders of Greek temples superimposed human anatomy over their blueprints to create what they considered ideal temples. Less visible human anatomy may have inspired some early architecture, according to New York surgeon Dr. J. William Littler, who compares the carpal arch of the hand (the unit of bones just forward of the wrist, anchoring the finger bones) to the Roman arch of Tiberius. The key to proportions for the Parthenon at Athens was supposedly based on the average height of a man, 1.681 meters ≈ 5′6″.

Early surveyors needed a unit of measurement that was portable, and

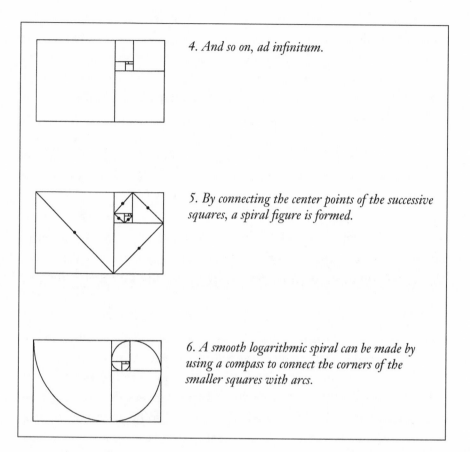

4. And so on, ad infinitum.

5. By connecting the center points of the successive squares, a spiral figure is formed.

6. A smooth logarithmic spiral can be made by using a compass to connect the corners of the smaller squares with arcs.

what more handy than one's own body, "the perfect invariant," wrote Paul Jacques Grillo in *Form, Function and Design*. The length of the arm from shoulder to finger tip inspired the unit of a yard, from elbow to the tip of the middle finger, a cubit. Among Egyptian pharaohs, the royal cubit was about a fifth longer than the common cubit of the little people, the taxpayers. The original fathom was equal to outstretched arms. In Japanese interiors, *tatami* units of 3 feet by 6, based on the dimensions of the adult human, arms outstretched, dictate the dimensions of rooms. The fingers of the hand are said to have golden proportions; the length of the longer bone is equal to the sum of the two smaller bones, which satisfies the Fibonacci sequence.

Surprisingly, Grillo says the unit of the foot came not from a human foot, but the distance between rungs of a ladder. The measurement was associated with energy in climbing, and ladder rungs were spaced a foot apart. The original measurement of an acre was also based on energy, the amount of land harvested or plowed in a day using primitive equipment. Grillo assigns some units of time to human energy; the 4-hour watch at the helm, he says, is also the limit of alertness for driving a car safely. The quart is based on the amount of liquid lifted with ease by one hand from the table, and the bushel, the amount of grain easily carried by one person.

Leonardo da Vinci used a measurement of the arm, or *braccio*, which was 1.91 feet long and which must have been the length of his own arm. The human body proved not to be the perfect invariant in Florence: the *braccio* was also listed as 23 inches for engineers and 21.7 inches for surveyors of land. In Rome, the *piede* was about a foot, or 11.73 inches, but in Milan it measured 17.134 inches. Leonardo collaborated with the monk Luca Pacioli on the book *De Divina Proportione*, which featured his often reproduced drawing of the nude male with arms tracing arcs, to demonstrate that armspread is directly proportional to height.

Every morning when Dr. J. William Littler dresses, he adjusts his tie clasp according to the Golden Mean. It is said to be an attractive proportion to humans, even to denote beauty in a face. A study by a psychologist named Buss mapped the ideal female face. The survey was based on the responses of 150 white males, and the fifty photographs of women included twenty-seven contestants for Miss Universe. The symmetry was measured in millimeters; a chin, for example, should be exactly one-fifth the height of the whole face. The Golden Mean has also been applied to the human face, with golden rectangles that have been considered to begin at the hair line and square off between lines drawn across the eyebrows, lips, and chin. The ideal face from Dr. Buss's study didn't exactly match the Golden Mean, but you could always find the numbers somewhere. As one author strained: "Amongst Miss Veronica Lake's measurements given in an American magazine, I notice 34" for the chest contour, and 21" for the waist, both Fibonaccian numbers, having a ratio of 1.619." Others describe links with the phrase "almost exactly."

In addition to being a "key" proportion of the human figure, the Parthenon at Athens is also credited for having proportions of the Golden Mean, which figures in the Great Pyramids of Cheops, the Temple of Artemis, the Cathedrals of Notre Dame and St. John the Divine, Stonehenge, playing cards, and business cards, but not the page on which these words are printed.

The golden mean has also been found in vases and canvas art. Often the proportion in landscapes divides sky from foreground, or in seascapes, sky from sea. Golden rectangles are superimposed over curving figures, or many figures. At times the connections are acute, because the artists themselves chose to follow the proportions and so divided their canvas; this was particularly true of the painter Piet Mondrian. For a logo for Manhattan's Riverside Park, the artist drafted first a Golden Mean, then began to play with the arcs. The wide use of the Golden Mean in art and architecture is credited as "the most elegant and efficient way of achieving unity among a diversity of elements." Greek columns have golden proportions; so do a pack of Marlboros.

A study of human adult navels came up with an average ratio of 1.618 to height, which brings to mind the man with his head in the oven, and his feet in the refrigerator, who was said to be pretty comfortable on average. I owe this chestnut to John Allen Paulos, whose lucid *Innumeracy* demonstrates how random events merely seem ordered. In a random computer printout of X's and O's, many people not only found patterns, but ways to explain them. Paulos's point was the value of distinguishing between coincidence and exclusivity.

Paul Jacques Grillo found evidence of the golden mean in the distribution of "major" planets within our solar system, *if* the distance of the earth to the sun is considered One. He suggests that such a "coincidence" (his quote marks) may be a clue to the evolution of the solar system, that current elliptical paths gelled from the original form of a log spiral. Galaxies do tend to form a log spiral, like Spiral nebula in Ursa Major. The so-called nebular hypothesis is old, a theory of Kant, Laplace, and others, proposing the solar system was once a nebula that condensed by centrifugal force.

The accumulation of sand grains on a spit (Koobi Fora, Cape Cod) also curve in a log spiral, as does a breaking wave. The form is widespread and ancient, shaping the tusks of mammoths, saber-toothed cats, and giant wart hogs, growing from one end, never changing its shape, accruing by logical additions. The coil of a nautilus shell or a sunflower grows by logical increments, and a species of caddis fly larvae builds nests of sand grains. The log spiral is the only curve that does not change its shape as it grows in size. "Nature does nothing in vain, and more is in vain when less will serve," wrote Isaac Newton.

Within a subdivided Golden Mean, the outline of a log spiral passes through the center of every golden rectangle within. Any section of the spiral is 0.618033... as large as the rest of the spiral. That the Golden Mean connects to the Fibonacci sequence was discovered by Jakob Bernoulli, the Swiss mathematician who in the late 1600s gave the name *Spiral Mirabilis* to the log spiral pattern. Bernoulli liked the form so much he asked that it be engraved on his tombstone, but was doomed to lie under the impression of a common coil.

D'Arcy Thompson points out that the original Fibonacci sequence has no *exclusive* relationship with the Golden Mean. He wrote: "Beginning with any two numbers whatsoever, we are led by successive summations toward one out of innumerable series of numbers whose ratios one to another converge to the Golden Mean." Instead of 1, 2, begin 1, 7. The sequence is then 1, 7, 8, 15, 23, 38, 61, 99, 160, 259, . . . , "and 99/160 = .618... and 259/160 = 1.619...; and so on."

Thompson draws a log spiral on gnomons, connecting the curves on a grid of tiny squares or hexagons. He suggests the true log spiral is found only in dead tissues, not the living. In a shell, "what actually grows is merely the lip of an orifice, where there is produced a ring of solid material," which he called the generating curve. The horns of kudu are "dead" cells; the same is true of the greater part of the nautilus shell, or a snail's shell. "In short, it is the shell which curves the snail, and not the snail which curves the shell." You could argue the reverse, and in

On Growth and Form, Thompson made a key point about the patterns of the living being no different from the dead, a concept not part of the conventional wisdom at the time. In the very next passage, "Spirals in Plants," Thompson notes that the primary shoots of plants are "analogous in a curious and instructive way to the equiangular spiral." Maybe the physical mechanics of growth were similar, but he was intent on opposing two beliefs, that numbers themselves were the force and that natural selection deserved any credit.

Thompson ruled out natural selection because spirals occur in many different climates and environments, "in the depths and on the shores of every sea," shaping many diverse creatures. These forms were recovered from "all periods of the world's geological history"; the shape of the nautilus is found among its fossil relics, and so is the shape of the pine cone.

Others still saw mystery. They drew a helix connecting a full whorl of leaves to a stem, and discovered a golden mean within the angles, described as ideal. "One irrational angle is as good as another," Thompson wrote. "There is no merit in any one of them, not even in the *ratio divina*," or Divine Ratio. " While the Fibonacci sequence stares us in the face" in the pine cone, he allowed, Thompson emphasized the opposition by Sachs, a botanist at Oxford, who considered it "a sort of geometrical or arithmetical playing with ideas" that were "gratuitously introduced into the plant."

Some of Thompson's critics said that he saw patterns where others didn't. His triangles were warped, his polygons were irregular, the hexagons he saw weren't perfect hexagons. "I believe that D'Arcy Thompson's critics were completely wrong," wrote Sir Peter Medawar. "Surely we must begin by seeking regularity. There is *something* in the fact that the atomic weights of the elements are so very nearly whole numbers; that there is a certain periodicity in the properties of the elements when they are written down in the order of their ascending weights; that hereditary factors tend to be inherited not singly but in groups equal in number to the number of pairs of chromosomes. Unless we see these regularities and strive to account for them, we shall not equip ourselves to understand or perhaps even to recognize the existence of isotopes, the significance of atomic number, or the phenomenon in genetics of 'cross-ing over.' We act either as D'Arcy does, simplifying and generalizing until the facts confute us. . . ."

"We want to know whether there is some single observation about the shapes of all plants and animals that makes sense, a statement that can explain how all organisms are different from non-living forms," wrote Stephen Wainwright of Duke; "There is such a statement: The bodies of multicellular plants and animals are cylindrical in shape." Three sen-tences later, Wainwright allows, "There are of course, exceptions—for example lettuce plants, box tortoises, sea urchins."

Wainwright also claimed that human designs are rarely cylindrical. Pens, pencils, water pipes, tail pipes, gas lines, drinking straws, gun bar-rels, telescopes, microscopes, kaleidoscopes, wires, cables, nails, pins, needles, trash cans, soda cans, beer cans, coffee cans, paint cans, ciga-rettes, cigars, sausages, clarinets, flutes, masts, booms and lifelines, tubes for vacuum cleaners, flag poles, broom handles, fluorescent light tubes, AA batteries, rolls of film, garden hoses, fire extinguishers, steering columns for cars, spokes for wheels, cylinders in car engines, fishing rods, Bonaventure hotels, silos, smoke stacks, tubes of lipstick, tubes of caulk, crack vials, bicycle pumps, chopsticks, cattle prods, thermome-ters, paper rollers on typewriters, batons, firecrackers, curtain rods, towel rods, chinning rods, rods you hang tents on, lightning rods, fishing rods, legs for tables, chairs, and tripods, drills for oil, necks for lamps, necks for umbrellas, walking canes, pea shooters, columns, vases, flasks, macaroni.

For every law, more than a few outlaws. For all the fruit blossoms that have pentagonal symmetry, there are tulips, lilies, and hyacinths that don't. A daisy may have five petals; so do we have five fingers, and the "blossom" on the test of a sand dollar has five "petals." But there is no golden mean to the sand dollar's test. Its shell is not a log spiral.

Yet there remains a yearning for a unifying principle, and when two or three patterns are linked, it is easy to be dazzled. "These connections may indeed be there," wrote Chris Williams in *Origins of Form*, "but one must be cautious, for these 'universal truths' are often so extraordinary

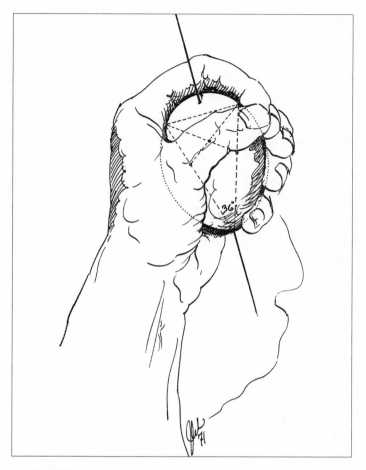

Dr. J. William Littler suggests that the human hand is composed of finger bones that represent the Fibonacci sequence, and the arc of the fingers, a log spiral. If an egg is held in this manner, it is said to be impossible to break by squeezing.

that they entice the faithful to push into realms where the law does not belong; for nature does work in many strange ways, not just one."

In Dr. Littler's paper "On the Adaptability of Man's Hand," he draws a hand embracing a chicken egg. The hand "adapts perfectly" to the curvature of the egg. The metacarpal arch conforms to the equiangular curve of the egg, distributing the forces of the palm and the fingers "so perfectly that even the most powerful grasp is unable to crush it." I did not believe this so I tried it, with the small end of the egg lined up at a tilt like the earth on its axis. I pressed gently at first. Then I pressed hard. Then I switched the egg to my right hand. The egg maintained its integrity.

Dr. Littler continues: "Assuming then a purposeful design in nature, one is compelled to look for certain structural and functional properties which can be expressed in simple mathematical terms. . . . The hand lends itself to such a study." The bones of a human thumb do not qualify for a Fibonacci sequence, but the fingers do, with the following sequence of measurements of the bone in centimeters, starting with the meta-carpal and working out to the finger tip: 8.8, 5.5, 3.3, 2.2. Dr. Littler said there is greater congenital deformation in the thumb than any other part of the human hand. "This shows that nature is still fiddling."

◆ ◆ ◆

IN *The Panda's Thumb*, Stephen Jay Gould suggests that ideal design is a lousy argument for evolution. The panda uses its thumb for raking leaves off the stems of bamboo; the wrist bone responded to new pres-sures, the radial sesamoid enlarged, and the thumb was constructed from available parts. Jury-rigged thumbs like the panda's prove that nature fiddles. A panda has the normal five digits beyond its irregular "thumb." "Odd arrangements and funny solutions are the proof of evolution— paths that a sensible God would never tread but that a natural process, constrained by history, follows perforce." Gould is careful to say "a nat-ural process" rather than natural selection, often credited as the only force of change.

Of course he buys natural selection as a powerful influence, and has explained Darwin into the bottom of the ninth. His column in *Natural History* provides a monthly refinement of Darwin's genius, twenty years running and counting. He has researched the Victorian naturalist to such an extent that phrases that slip by other readers are given piercing interpretations, based on the context that Gould has amassed. His research included Darwin's letters and lesser known works, some at Dar-win's home south of London, now a museum known as Down House.

In 1986 I visited this mansion with Professor Gould, and we took a walk on Darwin's thinking path, known as the Sandwalk, a circular promenade through a stand of trees that Darwin himself had planted in a remote corner of this 15-acre estate in Kent. It was here that he took daily walks to solve puzzles before returning to his study.

At the beginning of the Sandwalk, Gould spied a flint stone. Darwin began his walks by stacking a number of large flint stones at the starting point; with each lap, he kicked one away. By the time he kicked aside the last stone, his problem had to be solved. In this terrain, one thinks of Sherlock Holmes, who puffed away mysteries like a metronome, having designated it a three-pipe or a four-pipe problem. Darwin's method worked fine until local children, hiding behind the hedgerow, replaced the flints on the pile after he had walked by, expanding some issues unaccountably.

Devoting more time to a solution does not guarantee a more refined result, and the same is true of evolution. In the science of form, some hesitate to use the word optimal. Steve Vogel emphasizes that what we see may not be the final form, that evolution is ongoing; "There may be a better solution that an organism hasn't discovered, or can't easily discover." Even with constraints, nature can be wildy experimental, with multifunctional forms that lead to unforeseen serendipity, and the most respected scientists can be caught in awe of harmony and symmetry. "There is special pleasure in discovering that all this spatial harmony follows from one simple principle," wrote Harvard's Tom McMahon and Princeton's John Bonner with regard to the log spiral. And John Krebs of Oxford: "You only have to look around at things in nature, even bits of your own body, to see how incredibly beautifully designed they are to do the job they do."

In Thompson's essay "Aristotle the Naturalist," he described how Greeks looked on nature more with their minds than their hearts and put it in the background "while their foreground is filled with the affairs and actions and thoughts of men." But to Italians, including "Leonardo himself—the subordinated background is delicately traced and exquisitely beautiful; and sometimes we come to value it in the end more than all the rest of the composition. Deep down in the love of Nature, whether it be the sensual or intellectual kind . . . lie the roots of all our Natural Sciences."

Paleontologists love their fossils, Gould confesses love for his snails; molecular biologists love their molecules, primatologists love their primates, and physicists love God's formulas. Thompson loved geometry but recoiled at the "mysterious" link of the Fibonacci sequence to architecture and music and described such a link as "inexcusable Pythagorism."

A sphere was long held sacred; the circle, divine. Aristotle reckoned that God himself was spherical: "God has his existence in everything, uniquely and eternally, and . . . he is uniformly the same and round as a ball." The same uniformity was given to the earth, which Aristotle placed at the center of a universe of spheres. The sphere has a practical attribute that can be considered optimal, since it holds the most volume for its relatively small surface area. Limited surface area discourages the escape of heat, which is partly why the dome-shaped igloo is so efficient—massive domes are used for stations in Antarctica.

The symmetry of snowflakes relies on a blueprint based on the atomic structure of water. In a water molecule, the two atoms of hydrogen sit on the single oxygen atom at angles of 120 degrees. Consequently, ice forms a hexagonal lattice. The reason snowflakes vary in final form is a result of their roll and tumble down through different temperatures, and variations in water vapor. Unique conditions build on the original blueprint of six edges. It is as pretty a metaphor as any for the modifications that occur in the organic in time. Crystals such as pyrite, like the nautilus, maintain the same shape no matter their size, albeit cubic.

The 120 degree angle can also be found in parts of a dragonfly's wing, where circulatory veins embrace cells with a hexagonal network, a meshing that occurs in the embryonic stage of the wings. The polygons of tundra, lake beds, and ice have patches that are hexagonal. In the tundra, individual polygons can be ten to a hundred feet across, amalgamating to cover hundreds of square miles. The configuration is a result of extremely cold temperatures contracting the ground, creating fissures in the topsoil. During the spring thaw, water enters the cracks and then freezes, and the annual repetition of this process causes the formation of wedges. Dehydration of the soda lakes in Africa's Great Rift Valley leads to an accumulation of soda in similar fissures, and the same dehydrating effect occurs in the basins of fresh water lakes and patches of bare soil, with the wedges of clay curling, baked by the sun. These geometric patterns are the result of physical influences, and Thompson felt that similar influences were at work on the living: the presence or absence of water, changes in temperature, the creation of cracks, or connections when things were pressed together.

Since lakes dry up with some regularity, there was obviously some selection for creatures that could survive on land. Consequently, fish with both lungs and gills had an advantage and resulted in marine mammals like whales that surface to breathe. A clever (but paranoid) design would incorporate maximum options, like a Swiss Army knife, since evolution cannot anticipate the future. An optimal form today may not be an optimal form tommorow. Our appendix is being recalled. Like Mattheck's hazard beam, you wouldn't want to use that blueprint.

When D'Arcy Thompson cast his Cartesian nets across whole forms of fish, he left the reductionist approach to others, and preferred to consider the transformation as a whole. One problem with his approach was that it was only two dimensional, with grids placed over drawings that were merely outlines. J.S. Huxley came up with a formula to compare relative body parts. The formula is $y = bx^a$.

In a human body, for example, y can be armspread, and b height; a is 1, since armspread is directly proportional to height. The formula can be used to compare different proportions of different species if they are similar in form: fishes, antelopes, rodents, or, say, all bovids. Jeremy Rayner used this method to compare wing size to body size among bats and birds. The data are used to make a log-log graph to show skews, which involves a dangerous simplicity. To accommodate such a range as occurs in nature, the log-log graph "collapses the spacing between numbers when the numbers are large, and blows up the spacing when numbers are small," wrote McMahon and Bonner; "One could argue that this is a very natural, even a 'biological' way of dealing with information." But the authors issued several disclaimers, the first one pretty substantial. The allometric formula ignores the complexities and details in changes in form, such as a panda's thumb.

"Drawing an allometric plot and putting the exact numbers in an allometric formula do not in themselves explain the changes in proportions; they merely show that there is a change in proportions with a change in size." The authors add a final caution: "In many allometric plots, there is a wide scattering of data points, even though the scatter is deceptively minimized in log-log graphs. The result is that one often cannot be certain of either the exact slope or the value of b for the line, and even more seriously, one cannot be sure whether it is a straight line, a series of straight lines with well defined breaks, or a continuous curve."

The process seems tortured, as when Jay Hambidge saw forms like the nautilus as whirling squares. Hambidge, who wrote about dynamic symmetry in 1919, preferred the study of the human skeleton over a sunflower, because "the bone network of a man's frame is more stable than the skeleton of the plant's and, what is of much greater artistic importance, man is and always has been recognized as the most perfect type of design in nature."

Other sweeping biases included the "inferiority" of Chinese, Persian, and Gothic art to the work of Greeks and Egyptians, because Hambidge found dynamic symmetry in the latter. "In fact no design is possible without symmetry," he wrote. "The savage decorating his canoe or paddle, his pottery or his blanket, uses static symmetry unconsciously. . . . As civilization advances the artist becomes more or less conscious of the necessity for symmetry or . . . design." Picasso drew inspiration from primitive art, the log spiral figures in the pre-Celtic art of Ireland, and a circular living floor was discovered at Olduvai Gorge, the floor plan for thatched domes of the Turkana people, and in the bomas of the Maasai. While it is easy to romanticize people who spend their lives camping out, there is evidence that ancient Eskimos had a sophisticated knowledge of designs in nature.

"Knowing that musk ox horn is more flexible than caribou antler, they preferred it for making the side prongs of a fish spear," wrote Barry Lopez. "For a waterproof bag . . . they chose salmon skin. They selected the strong, translucent intestine of a bearded seal to make a window for a snow house—it would fold up for easy traveling and it would not frost up in cold weather. . . . Polar bear bone was used anywhere a stout, sharp point was required, because it was the hardest bone."

Their form of transport was built solely from organic parts. The sled was flexible, which allowed it to move over bumps on the ice without tipping, and over snow with lubricity. The runners were made from fresh char, wrapped and frozen, and the two units "cross-braced with lengths of caribou antler. . . . The bottoms of the runners were shod with a mixture of pulverized moss and water, built up in layers. On top of this peat shoeing came an ice glaze, carefully smoothed and shaved." They also made foul-weather gear from the guts of seals. "With a minimum of materials," Lopez writes, "historic Eskimos created a wealth of utilitarian implements, distinguished by ingenuity in design, specificity of purpose, and appropriateness of material to the task." In *Arctic Dreams*, Lopez also cites genius for repair: when confronted with broken engines imported by explorers, they fashioned a substitute part "with the right tensile strength or capacity for torsion or elasticity, something with the necessary resistence to heat, repeated freezing or corrosion, . . . a serviceable if not permanent solution. . . . Very sharp, some once said, these broadly smiling men with no pockets, no hats, and no wheels."

Alfred Russell Wallace argued that natural selection deserved credit for everything but the human brain, granted by divine intervention, which could explain the faulty connection between the brain and reproductive organs. According to Wallace, natural selection had a design plan. Every adaptation was a refinement. His followers replaced the role of the Creator with evolution, and reintroduced the mythical aspect of harmony and perfection.

The cleave to symmetry and perfection is tenacious, and cyclical, revived from the early phases of quantum physics and the Law of Least Action. "What satisfaction for the human spirit that, in contemplating

these laws which contain the principle of motion and of rest for all bodies in the universe, he [meaning us] finds the proof of existence of Him
who governs the world." This was the view of the French scientist Mau-
pertuis in 1744, partly inspired by the formula $E = \frac{1}{2} \times$ mass $\times$ (velocity)^2,
devised earlier by Gottfried Leibniz.

Leibniz is remembered for his 1710 essay "On the kindness of God,
the freedom of man, and the origin of evil," in which he tried to explain
how, in an imperfect world, God's holiness prevented him from prevent-
ing evil yet still allowed him to produce the best of all possible worlds.
Voltaire parodied the notion in *Candide*, which earned the book a place
on the Papal Index, then the short list for works of genius.

In Voltaire's satire, the student Candide listens to Dr. Pangloss, a
philosopher. "It is proved that things cannot be other than they are, for
since everything was made for a purpose, it follows that everything is
made for the best purpose. Observe: our noses were made to carry spec-
tacles." Contemporaries of Darwin, even the learned Richard Owen, said
that horses' teeth were arranged to carry the bits of a bridle. Dr. Pangloss
would have said that eggs were adapted to demonstrate the elegant grasp
of the human hand. (D'Arcy Thompson wrongly described the reason
for an egg's shape; he thought it had to do with the way they were
arranged in a nest, to keep them from rolling out, and he had the way
they barreled out of the female reversed. In the more classic question of
which came first, the chicken or the egg, since birds evolved from a form
of dinosaur, the egg wins. Baby dinosaurs have been found inside fossilized eggs, around long before the evolution of the chicken.)

Looking at the final form of things to explain their history prevailed
among evolutionary biologists in the U.S. and England for nearly a century. Recall Michael Foster's letter to D'Arcy Thompson, in which Foster pushed Thompson to explain how, in each case, natural selection was
at work, when Thompson was discerning physical forces. This attitude
inspired a critique of natural selection as the lone force.

In 1979, Steve Gould and Richard Lewontin blasted scientists who
hold "faith in the power of natural selection as an optimizing agent."
Readers got two titles for the price of one: "The Spandrels of San Marco

and the Panglossian Paradigm." San Marco refers to Saint Mark's cathedral in Venice—like most, it possesses a dome. Spandrels are the triangles that taper down when two arches meet at right angles. The mosaics applied follow the forms and radiate so beautifully that it is easy to think the spandrels were designed for the resulting images, when the opposite is true. The authors use this to demonstrate architecture as a constraint; arches are required to support the dome and the images are a by-product. The same analogy holds true for many structures: the ceiling of King's College Chapel in Cambridge, where the fan vaulted ceiling is embellished with the Tudor rose and portcullis. The exquisite colors of butterflies are a by-product of the architecture of their wings; tiny scales are part of the basic *Bauplan*—even dull moths have them. But with butterflies, the final trim visually overwhelms the basic structure. The petals of the sand dollar are merely a result of the exterior structure accommodating tube feet inside, not a flourish added for decoration.

Gould and Lewontin took a swipe at reductionists, but the thrust of their paper was to emphasize that natural selection is not the only source for change; mechanics and physical forces configure. In addition to crediting the insights of D'Arcy Thompson, they mentioned the work of Dolf Seilacher and his studies of architectural constraints, looking at organisms as modified blueprints. If natural selection could optimize parts separately, they conclude, "then an organism's integration counts for little." The tube feet of the sand dollar connect to the hydrology system and food groves of the poinsettia pattern on the bottom. It would be difficult to alter the architecture without affecting things in the interior. The Aristotle's lantern in sand dollars is much flatter than those in fully vaulted urchins. The enlargement of the bone in the panda's wrist affected the covering skin and attached muscles and circulatory system. And human anatomy is far from perfect. The pelvis, for example, is scarcely an optimal form, and neither is the human spine, both suffering from the adaptation for upright walking.

In the optimal foraging theory, John Krebs of Oxford and David Stephens of Amherst proposed a mathematical model meant to translate the efficiency of natural selection. Their focus was the economics employed by birds, "the average rate of energy gain per unit of foraging time." Birds not only conserve energy by soaring but also in their choice of food sources and the way they gather food. Hummingbirds expend great energy to cover more territory faster, but songbirds will abandon a diminishing food source for one that provides greater return. The mathematical model was hammered because it tried to quantify fitness, and bird behavior is shaped by many things, not just diet. Reproductive behavior and defense also figure, and what worked before is not always efficient in time. The foraging habits of the dodo, the extinct giant pigeon of Mauritius, were blamed for its demise, since the dodo was said to depend on the seeds of a certain tree that became extinct about the same time. The last dodo was seen in 1681, but surviving trees, less than a hundred years old, were listed in a 1941 survey, so the numbers didn't add up to cause and effect. The more likely cause for extinction was dodos' inability to fly, which made them more vulnerable to new predators, including settlers who brought along hogs; both the Dutch and their swine liked to eat dodo eggs.

From the log-log graph to the optimal mathematical model, numbers can be used to contort facts or fix them. Most of the time we get it half right. In *Gulliver's Travels*, Jonathan Swift parodied a court of mathematicians in Lilliput, with proportions that D'Arcy Thompson double checked. If Gulliver was twelve times their size, he would have collapsed from his own weight, increased to the third power, while the dimensions of his leg bones only increased by the second. Swift paid close attention to the "arithmetic of magnitude," Thompson noted, "but none to its physical aspect."

Behind numbers are concepts. With so many different forces shaping organisms, is it possible to develop a unified theory of form? Thompson pushed for physical forces, but they do not have the overarching effect he implied. Nor does natural selection. Life is complicated.

In the 1970s a unified theory called Supersymmetry emerged. It was the notion that particles were strings rather than points. Strings can vibrate, and the rate at which they vibrate suddenly mocked the lyre that

Leonardo da Vinci played. Even the theory seemed harmonic. "Maybe it isn't true," allowed Nobel physicist Steven Weinberg. "Maybe nature is fundamentally ugly, chaotic and complicated. But if it's like that, I want out."

People working in theoretical physics "often ask why nature chose math as its language," wrote Dick Teresi, co-author of *The God Particle*. "Why is it that the overarching principles of the universe can be broken down into equations? What the theorists are also saying, implicitly, is that they speak nature's language and the rest of us don't."

Some attempt to overcome this gap. When Murray Gell-Mann named six quarks up and down, charm and strange, top and bottom, it had a strange charm all its own. But when the team at Fermilab calculated that the top quark had a mass of 174 billion electron volts, give or take 17 billion, most of us had to wonder how many zeros that might be. The issue is not numbers, but the forces that generate them, the concept.

The experiment recreated a moment from the beginning of our world, the Big Bang. Ironically, William Broad wrote in the *New York Times*, "The discovery in all likelihood will never make a difference to everyday life, but it is a high intellectual achievement because the Standard Model, which it appears to validate, is central to understanding the nature of time, matter and the universe." Yet this understanding is what many people seek, and, as Dostoyevski said, ideas have consequences.

◆ ◆ ◆

FROM the thirteen books of Euclid's *Elements*:
A point has position, but no magnitude;
A line has position and length, but no width or breadth;
A straight line is a line which lies evenly with the points on itself;
A curve is a line of which no part is straight.
Book VII says: a perfect number is that which is equal to its own parts.

An imperfect number is deemed irrational. Ironically, the brotherhood that tried to corner the formula for eternal life denied the notion of infinity. Pythagoras had ideal numbers, whole integers without fractions. The most important numbers were five, the Pentad, and ten, the

Decad. Five was the number of love, uniting two (the female) and three (the male.) Seven was the virgin, since a circle cannot be divided into seven parts. Five was also the symbol of health and harmony, and it was used to compose the pentagram that was the secret symbol of the organization. Hippocrates was kicked out of the fraternity for exposing the pentagram. Hippasus, who published a drawing of the symbol, was supposedly deep-sixed during a storm at sea. The difference in penalties had to do with uttering the unutterable.

One of the Pythagorean theories stated that the sides of a right triangle are related by the formula $a^2 + b^2 = c^2$, where c is the longest side, and a and b meet at a right angle. This led them to investigate the square root of 2 and uncover irrational numbers. It was an affront to their entire universe, which relied only on whole numbers or clean fractions. There were no pi's. Apparently, it was a pledge among the pledges to keep this a secret, and Hippasus spoke of it. Keeping the truth at bay led to such political success that Pythagoreans gained power in the Magna Graecia, which included Sicily. Pythagoras, however, was forced to flee the country, and the leading members of his sect were besieged by a mob and perished in a fire. The pentagram survived, with a G in the middle to represent ten from a Latin translation of the Hebrew "yod" for ten.

The Pythagorean fraternity was established between 580 B.C. and 500 B.C., and became a model for the European monastic, and to some extent the Masonic lodge. The pentagram, with five stars linked, was the secret sign of the brotherhood. Pledges graduated through stages of novice, initiate, philosopher, studying the laws of numbers, and pledging an oath for the master, who addressed them from behind a cloth, like the Wizard of Oz. When they had achieved a state of moral purity as mathematicians, they were allowed to see Pythagoras face to face. They swore their allegiance on a perfect triangle of ten. The pentacle of their fraternity became associated with the occult: "the one who succeeds in linking usual and natural numbers to divine numbers will operate miracles through numbers." Consequently, crystals are thought to have healing powers because of their geometric form.

To some, the best miracle anyone could calculate was eternal life, and the Pythagoreans believed their souls would ascend to a meeting with God, passing through spheres, sacred circles that held the universe in tandem. The theory of heavenly bodies maintained an influence until the sixteenth century, when Johannes Kepler finally stopped forcing his data on the movement of planets into perfect circles. In 1619, Kepler wrote "Harmony of the World," praising the Lord for heavens that sing in ineffable language. To demonstrate this harmony, he devised an individual score for each planet.

The Pythagoreans got a lot of things right. They were the first to discover what Fibonacci rediscovered with his rabbits. The golden ratio was known to the neo-Pythagorean school, and described by Greek geometers of the Platonic school as "the section par excellence."

The Fibonacci sequence occurs in music. A score is drawn up from an eight-note octave. On a piano keyboard, this is produced by five black keys and eight white keys: 5, 8, 13. It's been suggested the major sixth chord is "the one our ears like best," by Randy Moore in "The Numbers of Life." "The note E vibrates at a ratio of 0.625000 to note C, only 0.006966 away from the golden mean. These notes produce good vibrations in our inner-ear's cochlea, a spiral-shaped organ. Every time I catch myself tapping my toes to the beat of my favorite song, I think of Leonardo the Blockhead, and how his 'numbers of life' describe what we like to build, look at, feel and hear."

Within our inner ear, the cochlea has a membrane described in *Gray's Anatomy* as "a remarkable arrangement of cells, which may be likened to the keyboard of a pianoforte." The spiral shape of the cochlea is of course the same as a conch shell, which resonates sounds around us that we don't normally hear. What sounds like rhythmic surf is a result of sound waves bouncing back and forth off the inner walls. Our eardrum vibrates at the same frequency as the sound waves. Nerve endings in the "pianoforte" cells translate the vibrations into electrical impulses, which is what a telephone and a microphone do.

For any sound, the slower the vibrations, the lower the frequency and pitch. When a pilot changes the pitch of a prop, the changing sound reflects the new angle of the blade. A lyre player alters pitch by changing the length of the vibrating string. The shorter a string, the higher the pitch, which is why children often sound like Lambchop— their vocal chords are shorter, and, generally, females' are shorter than males'. While humans can hear frequencies of about 20–20,000 vibrations per second, the range among musical instruments is narrow. A piano's highest note is only 4,000 vibrations per second. Their strings alone do not produce much sound; it is the passage over a wooden bridge that generates the sound.

Pythagoras worked out the first formula for tuning, which was applied to the piano for over a thousand years. According to his method, the second note of a scale should have a frequency 9/8 that of the first. The third note should have a frequency 10/8 (or 5/4) of the first. And so on. If you start the scale on the note middle C, with a frequency of 256 vibrations a second, then the next note, D, is pitched at 288 ($256 \times 9/8 = 288$) and the following note E is at 320 ($256 \times 10/8 = 320$). But if you started with D, E would be 324 instead of 320. It wasn't until the sixteenth century that a better formula was devised for tuning, to multiply a note's frequency by 1.059, with half notes compromised in between. Tempered tuning led to Johann Sebastian Bach's *Well-Tempered Clavier*, clavier meaning keyboard. The work consists of forty eight pieces, and lasts for about four hours. The eccentric WNCN radio host William Watson is remembered for his spontaneous decision, on a live broadcast, to play the whole thing again.

Resonance occurs when one vibrating thing sets another vibrating, and they share the same frequency. Molecular frequencies have been translated by David Deamer, a biophysicist at the University of California at Davis. An amateur musician known to play an accordion in Italian restaurants, Deamer noticed that the written sequences of DNA resemble musical riffs. The basics are called A, T, G, and C for the nucleotide base units adenine, thymine, guanine and cytosine. Deamer transposed the T's to E's, and suddenly the pattern sounded musical. He recorded his rendition of a snippet of a cow's DNA, called "Bovine Satellite," and people meditate to his slow and repetitive version of a human antibody.

With amplitude, resonance can cause vibrations of such a frequency that a singer can shatter a champagne glass. Power is not the issue, but the ability of the structure to withstand the increasing oscillations that have been set in motion. This was demonstrated by the bridge known as Galloping Gertie. Built in 1940 to span the Tacoma Narrows in Washington state, the bridge was not overly long at 2,800 feet. But the structure didn't have enough strength to resist torsion, and even the most gentle breezes made it sway and ripple. Just as there is frequency in sound waves, there is frequency in eddies of wind, swirling on one side, then the other of an object, which is what makes a flag ripple. The breeze that caused the eventual, dramatic collapse of Galloping Gertie was traveling at only 42 mph. Film footage of the collapse is often shown to engineering students; the same principles of rhythmic variations can topple trees when winds tease their individual harmonics into variations not inherent in their own range.

While Buckminster Fuller is remembered for his geodesic domes, he was also intrigued by the dynamics of eddies in the wind.

6. Tensegrity

I love you for the way you prognosticate.
— FRANK LLOYD WRIGHT to Buckminster Fuller,
in a 1938 letter meant to review Fuller's book
Nine Chains to the Moon

Saturday Review

"I KNEW everybody would call it a car," Bucky Fuller said, nailing an easy prediction, then described what would go down in history as a car as "the land taxiing phase of a wingless, twin orientable-jet stilts flying device." The jets were said to be inspired by a duck, and the general form, a fish. The car that emerged was streamlined and stunningly so, introducing aerodynamics to the road when Fords were shaped like shoeboxes. In a clever overview of evolutionary stasis in human designs, Fuller drew two blueprints with rectangles to outline a horse-drawn carriage and a 1932 Ford with the same outline, the engine a mere rectangular package of the four-legged one. He would hide his engine in the rear, a novel idea at the time.

Assembled in 1933, Fuller's experimental car had a nose with an uncanny resemblance to minivans introduced as New more than a half a century later, sleek and linear, with a windshield angled back at forty-five degrees, and the mien of a bottom feeder. Seen from above, the outline of the body was that of a long teardrop, from the side, a cigar. It was nineteen and a half feet long, and could carry ten passengers in addition to the driver. It was a stretch limo. It was a larval Land Rover in a meta-morphic cocoon. It was a three-wheeler.

It had front-wheel traction, and the third wheel at the rear was used for steering "the way a fish or a bird or a boat or an airplane must steer." This arrangement provided a maneuverability Fuller described as omni-directional, omni being a pet word of his, and the trajectory of his mind and myriad projects, which he deemed four-dimensional. Like his more famous geodesic domes, assembled in the blink of an eye, the car made quick work of a 180 degrees, pivotal to the chase. ("Like if the cops came after me, I could turn and go in the opposite direction. They could never catch me.") Bucky Fuller gave new definition to the shortest distance between two points (the gist of geodesic, lines on a sphere), a concept that prevailed with the car. You could parallel park by sliding sideways into a space. Nose into the curb, bring in the rear.

The line drawn by the rear wheel would not be perfectly straight, but a subtle arc. "Nature uses only waves—never straight lines," Fuller declared, overlooking crystals. His focus was on energy and movement in natural elements, when the shortest distance on a flight plan from London to New York rises like the St. Louis arch and includes a view of Greenland. Even people walking do not prefer a straight path, and when a sidewalk in a park is buried by snow, footprints create a meander. Cows, elephants, and antelope also meander, and only the smell of water drives a migrating herd of gnus in direct vectors on their approach, like the vectors of a sailor's tack.

The nautical influence of a steering rudder on the car was no acci-dent. Chief engineer on the project was naval architect W. Starling Burgess, who had designed two America's Cup winners. In a barter deal, Fuller helped Burgess build a sloop for the Bermuda race, indulging his own lifelong passion for boats begun on his grandmother's real estate in boatwright heaven, Bear Island in Penobscot Bay, ten miles off the coast of Maine. He grew up rowing to and fro from the island, and his first invention was a pole with a device that mocked local white jellyfish, opening and closing like an indecisive umbrella as he pushed and pulled it up and down, with the transition going wide at the water's surface to gain in propulsion, then shrinking to glide through the air. Air offers less resistance than water, as dolphins apparently learned, gaining speed as they porpoise. Fuller finessed this, with the bonus of being able to see where he was going, rather than rowing backwards.

During World War I, Fuller served in the Navy, a stint he recalled with unrestrained gratitude, embarrassed as he was to find himself having the time of his life when others were losing theirs. He saved a few lives by constructing a crane to pluck crashed sea planes out of the water. Throughout his career, he considered the structural integrity of ships superior because they survived natural calamities as a matter of course: floods (of the decks) and crashing waves, with an impact he compared to earthquakes and avalanches. Other attributes of the nautical influenced him, such as stingy habits to conserve fresh water, and the surprising economy of the petite on the offense. "The greatest secret of the Navy was doing more with less," he recalled, "where a little ship might be able to sink a big ship if it could move faster and outmaneuver it."

With speed and maneuverability in mind, it was also no accident that the shape of both the Bermuda sloop and the car was that of a long teardrop. You could superimpose the shape on many natural forms: a dolphin, the profile of a wing, a leaf, a sperm. "I am not trying to imitate nature," Fuller once said, "I am trying to find the principles she uses."

His ideas inspired Stewart Brand to publish *The Whole Earth Catalog;* another San Francisco entrepreneur, Lloyd Kahn, published *Domebook One* and *Two,* promoting "Economical and orderly use of materials" and "Minimum violation of the land."

Today this is known as eco-design, and in some cases it is too fashionable for words. In 1992, a New York architectural firm published a nine-point plan for sustainability, generally laudable, but they committed verbicide in point number six: "Eliminate the concept of waste." It's difficult to trim waste if you have no concept of what is wasteful. The abuse of semantics was a bane to Fuller, and led him to develop a lexicon. Waste was defined as the escape of energies from a preferred pattern. This gave it utility, like the car brake that Paul MacCready designed, capturing energy otherwise lost. Conserving energy included things like insulation, passive solar, and Fuller's clever use of reversible fans for cooling and heating his Dymaxion houses. This is what the other guys meant, but the articulation of architects seemed hinged by right angles, Fuller complained, deriding the conventional wisdom of "the solid, cubic diversion." His words took hold during a time when it was not cool to be square. Bucky Fuller peaked during the sixties and seventies and showed a consistent ability to render his listeners thunderstruck until he died in 1983. One of his admirers rightly predicted that Fuller's vision of utopia would die with him; the momentum was all in his rhetoric.

Even at the age of 81, Fuller remained unexhausted by this description as "geometer-mathematician/architect/designer/engineer/publisher/writer/inventor/songwriter (Roam Home to a Dome)/citizen of the cosmos," and described organic engineering as so superb "that our own complexedly-grasping fingers operate on the ends of extraordinarily powerful multi-jointed cranes responding exquisitely to every command of our brains, while, for instance, in my 81-year case, continually replacing all the working parts with approximately no lost operational moments." In works (often loving) that provided many quotes and details for this account, his life has been recounted in books with effusive titles: *An Autobiographical Monologue/Scenario*, for example, and before that, *A Spontaneous Autobiographical Disclosure*. Yet he was anything but spontaneous in his early 30s. In 1927 he stopped talking.

For nearly two years, Bucky Fuller held a moratorium on speech. This was difficult for his wife, since it coincided with his moratorium on making money. Wealth, Fuller decided, was energy directed by knowledge. He got the notion that some people should be paid not to work, a reverberation of the contemporary cliche in Russia: "We pretend to work, and they pretend to pay us." Fuller reckoned that a good number of people ought to be paid to think rather than simply show up and conform. "We should pay them to go to school, or to go fishing. Minds can marvelously repattern themselves while fishing." The innovations of a few would be greater than the losses for many, he figured. His outrageous idea was adopted in a discriminating manner by the John D. and Catherine MacArthur Foundation, which established the so-called genius awards. No strings are attached to the grants, but selection standards are high and include peer selection in secret. Fuller made not for profit a priority, and described it as his blind date with principle. Making a living was distinguished from getting a life, since both of his moratoriums resulted from his decision not to jump in Lake Michigan and drown.

He was drinking, he was depressed, and he stopped using words to regain his intuition, to hone the "multiplying nuances of my discoveries." He called the results "my experiential mathematics vocabulary."

"We begin by describing the shape of an object in the simple words of common speech; we end by defining it in the precise language of mathematics," D'Arcy Thompson wrote in his 1915 paper "Morphology and Mathematics." "Thus, for instance, the form of the earth, of a raindrop or a rainbow, the shape of a hanging chain, or the path of a stone thrown up into the air, may all be described, however inadequately, in common words; but when we have learned to comprehend and define the sphere, the catenary, or the parabola, we have made a wonderful and perhaps a manifold advance. . . . These words or symbols are so pregnant with meaning that the thought itself is economized; we are brought by means of it in touch with Galileo's aphorism, that 'the Book of Nature is written in characters of Geometry,'" a thought so pregnant that Thompson repeated it in *On Growth and Form*.

When Fuller found his voice, his phrases tumbled like rapids on the Zambezi. College students such as myself scribbled to keep pace with sentences that, like many during the late sixties, took trips without pattern or plan. Fuller's had a pattern that was lyrical if obtuse: "All mathematicians assume that a plurality of lines can passage through the same dimensional point at the same time, whereas physicists find that a physical-reality line is an energy event and that only one energy event can transit a given locus at a given time, all of which accounts for the physical interference patterns and angular reflections, refractions, and smash-up dispersions visibly occurring in their particle bombardment cloud chambers." His lexicon was more acute, perhaps because Hugh Kenner, one of his Boswells, distilled the meanings in a glossary. Geodesic is the word people remember; Dymaxion had the unfortunate ring of a brand name. The most powerful word in his lexicon was tensegrity (tensile + integrity) reflecting the lightweight strength found in the organic.

The work of Frank Lloyd Wright is described as organic, yet he rarely drew blueprints from nature in a functional sense. What Wright did best was hide his blueprints within nature, making structures unobtrusive. They seemed to belong on the landscape; they nestled and said Find me. Some were virtually hidden by trees, and balconies served as a trellis for vines. The coiling shell of the Guggenheim Museum is an exception to the above and everything. Wright certainly drew inspiration from nature and had a high regard for organic textures. For private estates, his materials were wood and local stone, bringing the outside inside, with spaces designed to confuse the difference. Skylights and windows were abundant, with patterns of ferns. Beams evoked the shadows of the forest. He loved oak trees, the way they spread, and his room plans spread, but not by fractals or the Baud curve. Beyond the Guggenheim and swimming pools, his forms were generally rectangular and linear, even rigid and suggesting bulk, and the only thing fluid was the water that might run through a structure, as with the Fallingwater estate in Pennsylvania.

Wright's use of white concrete has been categorized as plastic because concrete can assume curves. Still it is bulky and bright, and some of his boldly cantilevered balconies stick out like things you might hesitate to walk under. This is not meant as a swipe, but in terms of the focus of this book, there were precious few lightweight, tensile, elastic, or otherwise organic engineering principles employed. There was no tensegrity, no warp, no net, no teardrop or dome. If he adopted the organic line, it was that of minerals, crystals jutting. While there was one hexagonal house, in the majority of cases the Wright angle was the right angle. Rooftops sat flat like a wholesale display of levelers. And there was nothing temporary or mobile about Wright's structures, aspects that both Fuller and Frei Otto achieved. Wright's cantilevers might express freedom, but they couldn't move.

Architecture has been described as frozen music, and many interpretations reflect a snapshot view of life on earth. Donald Hoffman, writing about Wright, suggests "nothing was more difficult to attain than the spontaneous simplicity of nature, that sense of inevitableness typical of living things." In the long-running movie of evolution, natural forms are not spontaneously shaped, nor was any form inevitable, else all wings would be of the same length and proportion. The sense of simplicity is a surface impression, and to a great extent surface impressions are and were the job of architects. When the American architect Louis Sullivan promoted nature's eloquence of organization, his colleagues got into the eloquence rather than the organization.

Some followers of the Geomorphic school merely copied the frozen curves of dunes or waves or botany, or dumped the equivalent of a Native American mound to make a house earthy. In 1970, Le Corbusier put architects in the "unhappy state of retrogression" and engineers on a roll. "The engineer, inspired by the law of economy and governed by mathematical calculation, puts us in accord with universal law. He achieves harmony." Frank Lloyd Wright was wise enough to say that nature was not easy to read and inexhaustible in that sense—while others read high levels of testosterone. "Nature sustained Wright's manliness in architecture," Hoffman wrote, "that quality which set his work apart from the canons of Arts and Crafts taste, so pallid and polite. The difference was virtually one of gender."

"*Fin de siècle* morbid misogyny" came to Rhoda Koenig's mind when she considered the "repellent nature" of Antoni Gaudí's work. Another critic, reviewing a 1957 exhibit at New York's Museum of Modern Art (MOMA), wrote that the "bizarre eccentricities of Gaudí's architecture make any kind of appraisal nearly impossible. Gaudí's preoccupation with organic forms . . . is today inevitably seen against a background of psychoanalysis."

Antoni Gaudi i Cornet (1852–1926) was interested in the "geometry of warped surfaces." The general impression is that of a meltdown. One critic called it "drunken art." Some of his work—doorways shaped like shells, intricate wrought iron, elegant furniture and stairwells—is beautiful like a garden. The delicate legs of a chaise lounge intertwine like vines. Inspired by the sea, Gaudí's residential rooms look like the result of underground streams. Other constructions, for parks and facades, smack of the Watts Towers, a southern Californian folly of broken bottles and tiles. His most prominent work is pointed: Barcelona's Sagrada Familia cathedral was inspired by something he had seen in Africa ("such strange primitive forms"). They resemble mosque towers of mud in the western Sahel, or castles of clay built by termites. One of his chapels looks like the inside of a sand dollar, with parabolic arches and unequal struts, as if the ceiling and floor were in the process of coming unglued.

What Gaudi did brilliantly was develop a novel way to determine the inclination of loads in a structure. He used an elaborate 3-D model called a funicular. It was an elegant apparition, like a cathedral of macrame, hung upside down. Every point dangled sacks of bird shot attached by a cord. There were a lot of cords. For the Colonia Guell church in Barcelona, Gaudi played with the length of the cord and the weight of the bird shot for ten years. The final blueprint was drawn by turning a photograph of the funicular right side up, then cords and sacks were erased. What Gaudi discovered by using gravity was that columns could function at a warped angle, even bulbous at the center, that he could play with conventional lines.

This warp D'Arcy Thompson demonstrated with his Cartesian nets, a method also applied to skulls and bones of the pelvis. For his Park Guell, Gaudi built a retaining wall that leaned like a human femur. Right angles became endangered in his work. He poached them with precision, testing compression in a conventional fashion. At a local bookbinder's shop in Barcelona, Gaudi pressed stone, studying where the cracks began. His experiments were very similar to the method that experts like John Curry use to see how bone fractures, putting a bone between presses and laying on the loads. Gaudi's Casa Mila is known as the house of bones, because the columns resemble a human tibia, or shin bone.

What would we have gotten if Gaudi had gone into partnership with some big-time commercial architect? This was posed in an article for the *New York Times* by Paul Goldberger, who reviewed a work of Santiago Calatrava, the 43-year-old Spanish architect/engineer inspired

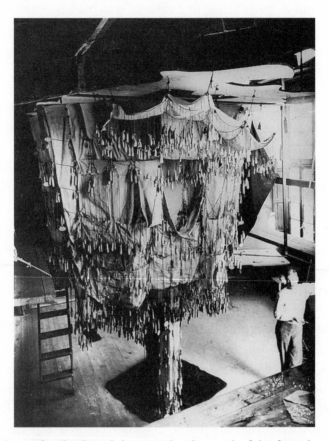

Architect Antoni Gaudí adjusted the many dangling cords of this funicular, which created cantenary curves like the drape of fabric, rather than sharp angles.

by anatomy, and trees, and who, coincidentally, was Catalonian, as was Gaudi.

In the summer of 1993, I stood in a ripe position to be swallowed by a Calatrava sculpture, *Machine for Making Shadows*, a motorized set of gargantuan white polished concrete ribs, which struck me as Moby Dick Breathing. I looked around the garden of MOMA for an old friend I had photographed twenty years earlier. Picasso's *The Goat* had a backbone and forehead reinforced by a palm branch that the artist had kept in his studio for a couple of years, wondering what to do with it. (Young Picasso had a studio in Barcelona, across the street from a mansion Gaudi designed.) The juxtaposition of ribs and palms continued when Calatrava designed a pavilion in Kuwait with moving roof ribs that resembled palm fronds.

Goldberger reviewed a galleria connecting two towers in Toronto; the big-time commercial architects were Skidmore, Owings & Merrill, based in the Big Apple. The structure was described as "nothing if not Gaudi-esque, a stunning six-story, 380-foot-long structure of white-painted steel and glass, leaping skyward with graceful, yearning arches." It was a Gothic greenhouse minus the vegetation, an omission which Calatrava overcame. The "challenge of the Sagrada Familia as a modern building" led him to design a Gothic greenhouse for the top of the largest cathedral in the U.S., Manhattan's St. John the Divine.

Seventy architects entered the competition to crown this cathedral, under construction for a full century and counting. It was intended that the finishing touch should demonstrate a spiritual link with ecology, and entries included disparate views—New Age and Dionysian—and a motorized garden that "spooked the jury." Calatrava's proposal of a 100,000-square-foot rain forest prevailed. It looked like trees and it had trees within it. It would be a biosphere. The notion was not new. Decades earlier, Buckminster Fuller had proposed a biosphere "as a central unifying tower" for St. John the Divine.

Calatrava said, "I wanted to create a parallel between the church and the bioshelter through the metaphor of trees: the roots are the crypt, the earth the nave, the transcept the trunk, and the pillars the branches.

The roof is leaves." Actually the roof will be a glass skin, inviting photo-synthesis.

It was in 1896 that Louis Sullivan wrote "form ever follows func-tion." Ironically, Sullivan's phrase appeared in an essay, "The Tall Office Building Artistically Considered," meant to defend skyscrapers. He argued that a building need not conceal its technology, that girders and beams should be exposed. The Seagram Building on Park Avenue added unnecessary girders to the outside to evoke this, in perfect misunderstanding, since their form has nothing to do with function and was wasteful to boot. During the late seventies, the High Tech movement exposed the guts of a building, including air conditioning vents, water pipes, and cables. They were painted bright colors to dis-tinguish their function, such that the Pompidou Center in Paris looks like a primer for Plumbing 101.

The architects for the Pompidou Center were not French but com-bined forces from Genoa and London. Piano & Rogers, young and rea-sonably astonished to win out of 681 designs submitted in 1969, were swept into a hastily arranged audience with President Georges Pompi-dou himself, who sat on a higher plane than his audience, one wearing a Mickey Mouse sweatshirt. Afterwards, they compared notes with the engineering team that had accompanied them, and everyone had noticed the same thing during the meeting. From their point of view in lower chairs, the soles of the President's shoes were polished. It was not an exterior gloss they meant to achieve.

"Sullivan's formula is really a moral proposition," wrote John Updike. "Form, in all aesthetic decency, should follow function—function and structure. But buildings, as more or less squarish lumps of shelter catering to human needs, cannot aspire to the naked mathematical beauty of bridges; one of their goals is the psychological comfort of their inhabitants."

Buildings are not so confined and include such unsquarish lumps as the A-frame and many round houses and lodges, residential domes and opera houses (such as in Sydney), concert halls (including a Fuller geo-desic dome assembled in 22 hours in Honolulu), sports arenas (the

Houston Astrodome), pavilions (Otto's Expo '67 in Montreal), restaurants (the Seattle Space Needle), and airline terminals (Eero Saarinen terminals at JFK and Dulles, which resemble a bird in flight, and a rolling surf, respectively). Round lodges and private houses imitate indigenous homes in Africa and Papua New Guinea. The Bonaventure Hotels in downtown Los Angeles and Atlanta are also round, with elevators of glass that expose their function in atriums. There are tents, and Mediterranean towers. Igloos. The Eiffel Tower. Gazebos. Greenhouses. The Crystal Palace. Pyramids. The Parthenon. Frank Lloyd Wright granted the Guggenheim the mathematical beauty of a coiling shell, and one of his disciples, E. Fay Jones, built a triangular cathedral of glass and timber in the Arkansas Ozarks. The Thorncrown Chapel, virtually transparent, is, to many, a bridge to heaven.

With regard to psychological comfort, Jones told a reporter from the *Times:* "A house ought to condition in a positive way. It should have a buoying effect. And if it somehow aligns itself with the attributes of nature, it might allow the residents to align themselves with those same natural forces of life." One of Jones's clients, an illustrator in Colorado, found his work improved by views of snow-capped mountains, as he worked beneath a ceiling of beams and glass. Novelist Ellen Gilchrist lives in a Jones house with a creek that runs through the living room, and my own office has a river running beneath it.

Buckminster Fuller also aspired to naked mathematical beauty, yet his smaller domes were surprisingly cozy in their interiors. To allay fears about living in an aluminum bubble, such as Fuller's first Dymaxion houses, a survey was conducted among housewives in Wichita, Kansas, who variously responded "It's beautiful," "I could clean it in half an hour," or "I want to buy it." Since twenty-six out of twenty-eight women had these responses, maybe the survey was multiple choice. In any case, *Fortune* magazine was enthusiastic. (Bucky Fuller was a consultant to *Fortune*, and wrote about technology and science: "The circular form, which arouses such doubts at first, looks quite unremarkable from inside and rather pleasant," a *Fortune* article said. "Most unexpected of all, perhaps, is the general impression of luxury.") Fuller's domes were appealing

Thorncrown Chapel, located in the Ozark mountains near Eureka Springs, Arkansas, is one of the designs of E. Fay Jones, a follower of Frank Lloyd Wright.

enough to sell by the thousands, and he and Mrs. Fuller lived in one. They were practical enough to appeal to the Pentagon. They were mobile.

Domes went to the equator, to the arctic, to the Persian Gulf. Even the concert hall dome in Hawaii was delivered (unassembled) by air. Transport and mobility were key concerns to R. Buckminster Fuller, and his Dymaxion dome houses drew from nomadic tradition, to travel light. Partly to emphasize their mobility, he priced his Dymaxion Dwelling units the same as a Cadillac, then around $6,500. The dome caught on as a dwelling for civilians much later, in the sixties and seventies, among the Volkswagen beetle generation.

Fuller liked to say that it took a quarter of a century from blueprint to the assembly line. "Many people thought I was some kind of nut," he relished, "because I was talking about air-conditioning, packaged [pre-fab] kitchens, and built-in furniture." In 1962, he discussed interactive TV, calling it two-way TV. "Children will be able to call up any kind of information they want about any subject and get the latest authoritative TV documentary. . . . The answers to their questions and probings will be the best information that is available up to that minute in history," a kind of superhighway. The Dymaxion car was developed decades before Fuller's success with geodesic domes.

In 1927, fifty years before Detroit went through its wing phase, Fuller envisioned a vehicle with collapsible wings, and a capability of "controlled-plummeting." It was the kind of feature that gives advertising copywriters a migraine ("air conditioning, aerodynamic design, and controlled plummeting"). This plummeting was inspired by Fuller's perception of ducks flapping; with every stroke, a duck "lifts and plummets, lifts and plummets." Fuller thought ducks were propelled by jets of air under each wing, but birds don't emit forced air like squids or nautiluses, which are truly jet propelled. Fuller gave squids credit, and his design for twin jets was deferred. Jet planes weren't refined until 1961. The propulsion for the car was a standard V-8, and the only real lift came from springs in the frame and a whitebread bounce because Fuller let some air out of the tires. But the teardrop shape was an effective airfoil, diminishing drag. The streamlining of the car was such that it reached 120 mph with a 90 horsepower engine, when it was estimated that the design of its contemporaries (kindly referred to as sedans) would require a 300 horsepower engine to attain such a speed.

For seven months, from January to July of 1933, engineer Starling Burgess and Bucky Fuller worked out of a building in Bridgeport, Connecticut, that belonged to a defunct factory with the dubiously succinct name of The Locomobile Company. Amazing to everyone but Fuller (who could stretch dollars as easily as he made them disappear) the project blossomed when the country was deep in the Depression; the car was assembled as FDR laid out plans for the New Deal. In addition to the innovation of front-wheel drive, the car had a body of aluminum, a chassis of steel, and a shatterproof windshield, one-eighth of an inch thick, adapted from a cockpit. Henry Ford gave Fuller parts at a 70 percent discount. (Fuller's first commissioned dome was for Ford headquarters in Dearborn, Michigan, and such an exception to his R & D mode that *Architectural Forum* ran a story, "Bucky Fuller Finds A Client.")

While no one claimed it could turn on a dime, the car did a whirligig down Fifth Avenue, describing a tight circle at many intersections from 57th Street to Washington Square. Fuller made his entrance at 57th by executing a sharp turn that appeared to be the work of Disney animators, which drew the attention of the traffic cop. "And while he was talking to me, looking in, I'd put my wheel completely over to the left, and go completely around him slowly, and he suddenly found himself facing the opposite way," A white glove went up on 56th Street, and then 55th Street, and so on. It wasn't a great day for mileage, normally 30 miles a gallon on an open stretch, but it was a good day for mileage in print.

Fuller's passengers included editors from the *New Yorker* and *Fortune*. A story by Archibald MacLeish in *Fortune* was read with interest by H.G. Wells in England, and during a 1934 visit, Wells was invited on a spin to Wall Street. A photographer from *Saturday Review* snapped him standing beside this timeless machine, said to have inspired the fictional vehicles in *The Shape of Things to Come*. To say that Fuller was a brilliant publicist is like saying Pavorotti shows promise. Fuller could have patented networking, too, simply by submitting his same drawing of the geodesic. A 1964 issue of *Time* had an illustrated cover with Fuller's cere-

The Dymaxion car had a single rear wheel, and front wheel traction. Buckminster Fuller, shown here in the driver's seat, worked on the 1933 aerodynamic design with nautical engineer Starling Burgess.

bral dome as a geodesic dome, a net of triangles. He thought globally, and showed no compunction about submitting his ideas on conserving energy for review by Leonid Brezhnev via then Canadian Prime Minister Pierre Trudeau, described as a friend. Anybody who was somebody in 1934 was given a lift in his Dymaxion car, including Amelia Earhart. Einstein doubtless would have gone for a ride, if only he had had the time.

According to *An Autobiographical Monologue/Scenario*, Fuller met with Einstein on Riverside Drive, the bastion of intelligentsia in the U.S., shortly after the physicist immigrated to the U.S. in 1934. The purpose of the meeting was to fact-check parts of Fuller's book, *Nine Chains to the Moon*. In it, Fuller entertained practical uses for the theory of relativity. This was said to amaze Einstein, who expected his work to be useful only to astrophysicists pondering. "I'm the only person that heard Einstein say that extraordinary thing," Fuller told his son-in-law, Robert Snyder, and kept a copy of Einstein's letter that led to the Manhattan Project.

Fuller's car had practical applications, although he said the prototype should be driven only by aviators. "She was the most stable car in history," he claimed, with a low center of gravity, and she did not skid, he emphasized, even during an ice storm "when you had to be careful not to slide off the road just standing still." These remarks came after a tragic accident, both for the people involved and the future of the car. On October 28, 1933, the *New York Times* ran a front-page story below the fold: "Upset of 3-Wheeled Car Kills 1, Hurts 2; Noted British Aviator a Victim in Chicago."

At the wheel was a race car driver named Francis T. Turner, hired by the Gulf Oil Company, which had purchased the car to advertise aviation fuel. A second Dymaxion car had been commissioned by a British group, who had sent aviator William Francis Forbes-Sempill over on the *Graf Zeppelin* for a test drive. The other passenger was Charles Dollfuss, of the French Air Ministry. The driver was killed, the distinguished passengers injured, and the car turned turtle at the most inauspicious place on the globe, near the entrance of the Chicago World's Fair. The brief, six-paragraph story in the *Times* reported: "The Zeppelin passengers were being rushed by Turner to the municipal airport to catch an airplane for Akron, there to rejoin their ocean-flying companions, when the machine skidded, turned turtle and rolled over several times. Police said it apparently struck a 'wave' in the road."

Fuller said his car had been rammed by a politician's car, which left the scene, but as Mark Twain put it, "A lie gets halfway around the world before the truth has time to put on its boots." The coroner's inquest was postponed for thirty days until Colonel Forbes-Sempill recovered enough to testify. It emerged that a second car owned by a Chicago South Park Commissioner *was* removed before reporters arrived. Apparently the two cars collided, after racing and weaving in and out of traffic at about 70 mph. In his book on Fuller's work, Dr. Robert W. Marks writes that by the time the facts emerged, the smashup had "lost its news value. The earlier reporting was never amended in the press."

Ironically, the car was repaired and sold to a director of the automotive division of the U.S. Bureau of Standards in Washington, D.C. While

Fuller could find nothing about the wrecked car to suggest any design failure, as a result of its nautical influence, it had one characteristic of a sail boat. When hit with cross winds, it tended to nose into them, instead of yielding like a normal car.

The British got cold feet, and the Dymaxion they ordered was eventually sold to a group of mechanics in Bridgeport when Fuller was strapped for cash. A third Dymaxion was sold to conductor Leopold Stokowski and his wife, Evangeline Johnson. It was put on exhibit at the Chicago World's Fair, part of Fuller's attempt to overcome the bad publicity. He spent his savings to build Dymaxion cars two and three. Over the next decade, car three was resold several times, then disappeared, to be rediscovered in a garage in Kansas, a state notorious for cross winds. Fuller arranged to get it back from the Wichita Cadillac dealer, and estimated its mileage at 300,000. Dymaxion car one was destroyed in a fire in a Washington, D.C., garage. Dymaxion four, designed to average 50 miles a gallon, never went further than the drawing board. The Dymaxion car was evidence of Fuller's ability to prognosticate; one of the best ways to predict the future is to design it.

Fuller built Dymaxion houses, Dymaxion bathrooms, Dymaxion Deployment Units, Dymaxion Dwelling Machines, Dymaxion catamarans, and designed a Dymaxion map, unfurling the globe in the first accurate rendition of the earth and its oceans on a single plane. The word Dymaxion has been credited to Waldo Warren, an advertising copywriter for Marshall Field in Chicago. The department store used Fuller's architecture to promote their furniture, but 4-D failed as a buzz word. It sounded like an apartment address or a bra size. Warren interviewed Fuller about his concepts and seized upon the words Dynamic, Maximum and Ions. In the Fuller lexicon, Dymaxion is defined as maximum gain from minimum energy.

Fuller was a pioneer at trimming waste in design, and credited the aircraft industry with ten times the efficiency of the automobile industry, based on "performance realizations as ratioed to invested resources." Consequently, he used the material and structural tricks of aviation, making everything lightweight. His Dymaxion dome shelters, designed to accommodate thirty marines, were delivered by a Sikorsky copter. Working with Beechcraft in Wichita, Fuller used the same assembly line that produced the fuselages of B-29s. His long-range notion was to provide a postwar conversion industry in housing, a ripe idea post cold war, as Boeing contemplates producing cars. It is one of his many ideas that ought to be dusted off and entertained. As Hugh Kenner wrote in *Bucky, A Guided Tour of Buckminster Fuller,* "Again and again, the newest experiences men can devise correspond to the oldest they can recover," a sentiment worthy of D'Arcy Thompson.

When Fuller entered the arena of architecture, he imported old ideas. "It hit me hard to find that the building world was thousands of years behind the art of designing ships of the sea and air," a theme repeated by Professor J.E. Gordon, in *Structures, Or Why Things Don't Fall Down.* Fuller's efficient approach to construction prevailed with the aluminum dome that crowned Ford headquarters in 1953. Assembled in a mere thirty working days, it weighed 8.5 tons; a conventional dome of steel would have weighed 160 tons. His Dymaxion bathrooms looked like airplane lavatories. The entire unit was prefab; the toilet was dry, with a storage unit for compost, and the lavatory and shower basin designed to recycle gray water. "I find humanity being incredibly careless and thoughtless in using water," Fuller wrote. "Human beings go to the toilet and get rid of one pint of liquid by using four gallons of water." The Dymaxion bathroom was patented in 1937. In 1990, the average daily flushes for one person in the U.S. consumed 30 gallons. Putting a sealed jug in the water tank is still promoted as the solution.

Fuller's regard for natural resources was such that he began a comprehensive survey: The Inventory of World Resources, Human Trends and Needs. It was not the first time the concept of sustainability was put into numbers. In 1798 Thomas Malthus published his "Essay on Population," which emphasized that organisms, including humans, can reproduce at a geometric rate, while food sources increase only arithmetically. Malthus predicted that poor people would suffer most, leading to conflicts between classes, clans, and countries. The predictability of his theory remains debatable, even during the U.N.'s 1994 conference on population in Cairo. Many of the delegates considered the kind of data that Fuller began to assemble years ago.

Fuller took Malthus's survey a step further, folding natural resources such as fossil fuels and minerals into his statistics. He also added recent inventions Malthus did not anticipate. Refrigeration, for example, didn't exist beyond ice boxes in 1798; the cross-country trucking of vegetables expanded distribution. Fuller proposed ingenious networks not only for food, but also for utilities, metals, and oil. His inventory had a scientific vitality, with numbers maintained and updated for fifty years, beginning in 1917. It was also graphic, with population clusters marked on a giant Dymaxion map.

From this big picture, Fuller devised the World Game. It was a peaceful counterpoint to war games, with the goal to remove conflicts over resources. The World Game was the flip side of Monopoly, and could not by any stretch of the imagination be called Isolation. Players were supposed to invent economic solutions in design, distribution, and energy. Fuller proposed capturing the emissions from smokestacks, and recycling the sulphur from sulphur dioxide. He also claimed that 95 percent of petroleum used was wasted. In his view, there was no energy crisis, but a crisis of ignorance.

The first World Game was played by some of his students in New York in 1969, and it resulted in a proposal to extend electrical power across remote regions of Russia and China. It would double the generating capacity overnight, Fuller said. Subsequent World Games developed the idea of harnessing energy from coastal tides and windmills. A 1970 project in Wisconsin known as Windworks led to direct consumption by utility companies in twenty-two states. Capturing tidal energy has proven more elusive, but experiments are currently under way on the coast of England.

Capturing wind for power is hardly a new notion; the Chinese accomplished this in the thirteenth century. Chinese inventors also held a high regard for the strength of the triangle, which Fuller recalled discovering during a session in kindergarten, when he was four and a half. When the teacher distributed toothpicks and peas, the other kids began to build square houses. Fuller had poor eyesight, and said he saw only amorphous shapes, with no sense of structural lines common to the landscape. He poked his toothpicks into the peas until his construction "felt good" and didn't warp when he twisted it. He said he felt what he would expound in later years: Triangles hold their shape, squares and cubes do not.

Synergy is the behavior of a whole unpredicted by its parts, and Fuller felt that humans could intuit this. "Synergetics arouses human awareness of the always and only co-occurring, non-tuned in cosmic complementations of our only from moment to moment systematically tuned-in conceptionings." He credited this awareness to some of his discoveries, being a bug in the big Brownian, a part of the universe, in tune with forces. He saw ways to bridge *The Two Cultures* of scientists and nonscientists described as unbridgeable by C.P. Snow, and met with Lord Snow over tea in London to argue his case. Science, Fuller said, had been on autopilot for too long, with no direct, tangible feel for navigation. The problem was semantics, the foreign language of science, noise that interfered with a grasp of kitchen science. Fuller liked to describe how he had realized engineering principles simply by handling an oar.

All things in nature, he emphasized, combine tension and compression. "However much we may find a tendency, whether in Nature or art, to separate these two constituent factors of tension and compression," D'Arcy Thompson wrote, "we cannot do so completely; and accordingly the engineer seeks for a material which shall, as nearly as possible, offer equal resistance to both kinds of strain."

Fuller demonstrated this with a rope, stretching it to show the tensile forces, yet the fibers compressed as the rope contracted. "Many architects mistakenly talk about using tension all by itself—but there's always compression occurring at 90 degrees to the tension." From ropes he moved to steel rods; when loaded from the top, they spread, or bulge. "As it gets fatter, its girth goes into tension. So while I am purposely loading—compression—it goes into tension at 90 degrees again. . . . If I keep on loading such a column, it will become a sphere."

◆ ◆ ◆

THE genius of the geodesic dome is that its strength increases with size. A bridge, merely doubled, is weakened by its own weight. With a geodesic dome, parts under compression (aluminum tubes) are suspended between tensile forces (thin metal wires), vectors that work like the spokes of a bicycle wheel to suspend loads. Fuller referred to this arrangement as islands of compression in a sea of tension. With a graduation in size, the islands diminish in number, or frequency.

Fuller's first geodesic dome was called an Octet Truss, and it was composed of alternating tetrahedrons (four-faced polyhedrons) and octahedrons (twelve-faced polyhedrons). The Octet Truss had Vector Equilibrium, the same equilibrium that occurs when hexagons interlock in a honeycomb, vectors being their angles of attachment. Hexagons alone cannot enclose a sphere; it is a mathematical impossibility known since Leonhard Euler pointed it out in the eighteenth century. Pentagons could be used as a link, but the economy and strength of two triangles, butt to butt, was Fuller's solution. For larger domes, he added icosahedrons, with twenty faces, increasing the frequency of vectors and tensile strength.

Over 300,000 geodesic domes were built during Fuller's lifetime, from a gargantuan envelope near Baton Rouge, Louisiana, that covers petroleum storage tanks, to pavilions in Kabul, Afghanistan, and in Moscow, where Khrushchev called him R. Westminister Fuller. Buckminster Fuller liked to say that his domes covered more earth than any other structure. He tried to include part of Manhattan, proposing a massive greenhouse that would have expanded high over the trees of Central Park. Because it is spherical, a dome doubled in size only increases its surface area by four times, but the volume increases eight times. The energy and heat of the city would be maintained in winter, he said, with no need for snow plows. What would happen in August is unclear from the works I read; presumably panels could be opened, but the notion of a Department of Thermostat Control is foreboding. The Big Apple dome would have been a mile in diameter.

Fuller first published his formulas for a geodesic structure in 1944; he wrote about the principles of the concept in 1933, but his interest in the problem is traced back to the same year *On Growth and Form* was published.

In *The Dymaxion World of Buckminster Fuller*, Robert Marks wrote:

> *Beginning in 1917, Fuller attempted to organize into a logical system the energy patterns he observed or discovered. The consequence has been an extensive series of propositions and demonstrations which, as a collection, he calls Energetic-Synergetic Geometry. What follows is startling. . . . Fuller became alert to the fact that there was a regularity of patterning linking the behavior of man-devised structures, such as bridges, buildings, frames, trusses, and the behavior of the minute or invisible structures, such as crystals, molecules, atoms. It seemed apparent to him that the patternings of force in a macrocosm were not essentially different from those in a microcosm; forces interacted in the same way, moving most economically toward equilibrium. . . . If this assumption were correct, Fuller reasoned, he could isolate in one coherent mathematical system the significant rules which govern all physical structures. . . . One phase of Fuller's exploration for a geometry of energy resulted in the discovery of what he named closet packing of spheres, each sphere being conceived as an idealized model of a field of energy in which all forces are in equilibrium, and whose vectors, consequently, are identical in length and in angular relationships.*

D'Arcy Thompson had covered the same ground, drawing bridge trusses that resembled skeletons, and describing the "Close Packing of Cells," a heading in chapter four of his book. His description of the common patterns of hexagons included the "equilibrium" that is achieved as a "condition of minimum potential energy," and he emphasized economy by citing the Law of Minimal Areas. He, too, measured the angles, or vectors, that are identical in what Fuller called closet packing of spheres. The whole point of his book was that fundamental rules govern all physical structures, and mathematical formulas had already been devised to explain the way forces move economically towards equilibrium. Thompson referred to the studies of several observers, including Ernest Lamarle, a Belgian who wrote about fluid films, or soap bubbles, in 1864, and Felix Plateau, who wrote on the same subject in 1873.

Thompson also described the basic building block of a geodesic dome, the tetrahedron.

"When Plateau made the wire framework of a regular tetrahedron and dipped it in soap solution," Thompson wrote, "he obtained in an instant a beautifully symmetrical system of six films, meeting three by three in four edges, and these four edges running from the corners of the figure to the center of its symmetry." The tetrahedron appears in the skeleton of silica called Callimetra, which Thompson included as an illustration, and described as a "*spherical tetrahedron*" [his italics]. He included other geodesic structures, Spirogyra and Radiolaria, in a short chapter, "A Parenthetic Note on Geodesics."

Fuller patented the geodesic dome in June of 1954, but according to Tony Rothman in *Science à la Mode*, the world's first geodesic dome was opened to the public in August of 1923. The dome was located in Jena, Germany, and sat atop the Carl Zeiss Optical Company. Chief engineer of Zeiss was Walter Bauersfeld, who designed the dome as a planetarium. The light iron framework was covered with a thin shell of concrete. The dome, called the Wonder of Jena, was destroyed during the Second World War, but Rothman obtained a copy of a 1925 patent from the construction company. The patent apparently did not detail a unique formula for the framework, but a "method for the fabrication of domes and other curved surfaces of reinforced concrete." The patent seemed to focus on the precise application of the concrete, with a thickness based on the ratio of an eggshell to an egg. The Zeiss Company built other domes, none of them geodesic.

"The German planetarium designer seems not to have known what he stumbled on, since his Geodesic Dome had no successor, and was attended by no explanations," wrote Hugh Kenner in his book on Fuller. "Can a man have invented such a thing, if he cannot explain it? The Patent Office avoids such metaphysics; all you can patent is a special-case realization."

Fuller entered his own defense in his 1975 book *Synergetics:* "But what often seems to the individual to be an invention . . . time and again turns out to have been previously discovered when patent applications are filed and the search for prior patents begins. Sometimes dozens, sometimes hundreds, of patents will be found to have been issued, or applied for, covering the same idea. This simultaneity of inventing manifests a forward-rolling wave of logical exploration. . . . I recall now that when I first started making mathematical discoveries, years ago, my acquaintances would often say, 'Didn't you know that Democritus made that discovery and said just what you are saying 2,000 years ago?' . . . Rather than feeling dismayed, I was elated to discover that, operating on my own, I was able to come out with the same conclusion of so great a mind. . . . Such events increased my confidence in the resourcefulness and integrity of human thought purely pursued and based on personal experience."

The formulas that Fuller devised are considered, by Robert Marks, a "mathematical discovery that is significantly and uniquely Fuller's." Fuller's patent prevailed, and the geodesic dome became his private postmark. He established two corporations (wholly owned by him) to deal with the licenses and accounting: Geodesics, Inc. for military and government use, and Synergetics, Inc., for private industry. Some companies "tried to sidestep Fuller's patent and found that evasion impossible," wrote Marks.

Since D'Arcy Thompson's book drew much from the history of math and geometry and biology, that same history was available to Fuller. D'Arcy Thompson was modest about his own contributions and generously credited the people who paved the way. Fuller, on the other hand, said that "operating on my own," he made "mathematical discoveries," and perhaps he did. That his interest in mathematics dates to the very same year that Thompson published his book could be contorted to accommodate a new theory by British biologist Rupert Sheldrake, that knowledge and ideas are in the air in global fields. Fuller claimed this.

One of Fuller's themes, according to Hugh Kenner, was that the mind was not local, "but integral, whenever it entertains principle, . . . that we are so designed that we can harmonize our decisions with the rest of the Universe. . . ." Many ideas that D'Arcy Thompson pulled together were in Aristotle's and Leonardo's atmosphere, and during Thompson's lifetime the field of ideas over Edinburgh never rained on Cambridge or Oxford.

The geodesic pattern and contortions of the triangle are common visions to migraine sufferers, who describe things that resemble Radiolaria, Callimetra, and Spirogyra, forms that have been around hundreds of millions of years, since the Cambrian. There may be absolutely nothing new under the sun. However Buckminster Fuller came upon his notions, he deserves credit for applying these blueprints to technology and architecture and communicating the concepts to a broad audience. His popularity and fame may have been the true source of irritation among his detractors, along with his hold on the patent. But critics did not buy the notion that he intuited what had required considerable R & D time by others.

Tetrahedrons of silk were used by the Chinese, who built the box kite over 2,000 years ago. In 1903, Alexander Graham Bell wrote, "This form seemed to give maximum strength with minimum material." As Philip Drew described in an essay on "The Lightweight Aesthetic," Bell invented several tetrahedral structures, employing prefab while Bucky Fuller was still in elementary school. Ten-inch mass-produced tetrahedron units were covered in silk, and one structure contained over 3,000 cells. Called the Cygnet, it was meant to be a flying machine. In 1902, Bell used his tetrahedron factory to build a temporary shelter for the livestock on his estate, and in 1907 constructed an 80-foot tower using vectors of iron pipe half an inch thick.

Drew, a former editor of the *Architectural Press* in London, also traces the structural evolution of the geotetic form. The interior network structure for zeppelins, for example, was composed of tiny triangles in a strut, a structure refined by Barnes Wallis. Wallis worked on the R100 airships from 1920 to 1934, developing what was called a geotetic principle. The struts were a lattice of holes, and in addition to providing lightweight strength for airships the design was used for hangars and the framework of a Wellington bomber in World War II. "It is worth noting," Drew writes, "that Barnes Wallis's enunciation of the geotetic principle anticipated by a quarter of a century Buckminster Fuller's appropriation of the idea in his geodesic domes."

The geotetic principle is a homogeneous distribution of stress, like the constant stress axiom that Claus Mattheck found in trees. "The great strength and extreme lightness of Barnes Wallis's geotetic structure arises because the structural system responds as a single organism," Drew wrote, "allowing the loads to be spontaneously diffused." Like a spider's web, a dome is also a lattice work that responds throughout when a single point is touched. This is the gist of tensegrity.

At the Institute for Lightweight Structures, Otto led a series of experiments with models, a process he calls Form Finding. During the fifties and sixties, the forms he liked to find were prestressed membranes. In 1951, while still a college student, Otto became intrigued by suspended roofs and was encouraged by Eero Saarinen. Like Gaudi, his students played with catenary nets, which were inverted to create lattice domes. Like Plateau, they dipped wire or strings into soap film, then twisted the strings to experiment with minimum surfaces. In 1964, one of Otto's students, Larry Medlin, discovered an innovative form.

Imagine a delicate lasso dipped into a bubble, to create an elliptical noose. The noose is then lifted by a cord, and the large bubble hangs like a deformed tent. The elliptical noose itself is suspended and supportive at the same time. This configuration was enlarged and stabilized with net wire membrane to create the German pavilion at Expo '67 in Montreal. The first experimental structure, secured by a dozen cables, and lifted by a 56-foot mast, serves as the studio for the Institute for Lightweight Structures on the Vaihingen campus near Stuttgart. It looks like a vast tent that could fly, not only because of the wire lattice but because the structure is mounted on a first floor of glass windows. At night, you can see the glow of ignition. There are no dividing walls within; the space is open. Students on one project work alongside others solving different problems. Otto's office was elevated on the second level, but he did not confine his influence to the center, and served as a visiting professor at Berkeley, Yale, Harvard, and MIT before being named an honorary fellow of the American Institute of Architects in 1968. Otto retired in 1991.

While Otto's regard for the tensile strength of spider webs has been

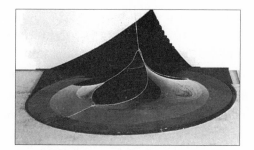

The soap film membrane, with a loop that lifted the structure to a mast, was the model for the 1967 German Pavilion and the Institute for Lightweight Structures, near Stuttgart, Germany.

mentioned, his fish-belly designs for bridges during the Second World War also employed tensegrity. He found that the volume of materials was connected to the number of tension members. Increasing the tension concentrated the compression in fewer short struts—islands of compression. Otto also found tensegrity in umbrellalike constructions and in lattice domes. In 1962, in Berlin, biologist J.G. Helmcke showed Otto some photographs of diatoms, small aquatic creatures with skeletons of silica. There are over 10,000 species of diatoms. Some have a lattice structure that resembles the botanical, as in Queen Anne's lace. Others are encased by a rim, with the close-packed cells creating a dome shell. Otto saw a diagram of forces that reminded him of his own blueprints for domes.

In his conversations with Helmcke, Otto learned about the evolutionary history of diatoms and later wrote: "In the evolution of living creatures, it is possible to see how redundant features and attributes disappeared or how sudden changes (mutations) . . . brought about lasting improvements in the efficiency of the organism. . . . It is conceivable that man, with his resources, can make the efficiency of his structures superior. . . . However, when we examine the structures of living creatures with regard to their efficiency, we cannot escape the conclusion that so far, man has hardly achieved anything like it."

Diatoms became the symbol for SFB 230. Their form figures on every document and brochure. By coincidence, the diatoms are among the forms that the sand dollar consumes. They are what we would consider high-protein food, providing more energy to the sand dollar than the food bits it trims off the edges of sand grains alone.

Otto's first glance at a diatom recalls the anecdote in *On Growth and Form*, when the engineer Cullmann takes one look at the lines on a bone, and exclaims, "That's my crane!" The dialogue between the architect Otto and the biologist Helmcke led to a joint lecture series at the Technical University of Berlin in the winter of 1962. Helmcke discussed organic shells, Otto, shells made by humans. This lecture series led to a symposium on Biology and Building, which evolved into the SFB research projects and annual conference, SFB 230. The value of such cross-disciplinary work was also of use to science, Otto noted in a 1962 article: "In recent years many initial steps have been taken in this direction, e.g., by R. B. Fuller, whose structural theories have led to a better understanding of certain aspects of virus research."

At an international conference of molecular biologists in February of 1962 a professor from Cambridge "announced" the mathematical principles that shape the shell of virus proteins. The *International Herald Tribune* reported "all these virus structures had proved to be geodesic spheres of various frequencies" and gave Fuller credit for discovering the principle in 1933. In 1958, the director of Nuclear Research at Dow Chemical, John J. Grebe, published a paper on fundamental particles. Grebe considered the mass of subatomic particles "highly reminiscent of a relation pointed out some years ago by R. B. Fuller." In 1959, two British researchers noticed a similarity between Fuller's geodesic dome and the structure of the polio virus.

The geodesic pattern also appears in one of the first multicellular creatures to appear in the ancient ooze, in the form of bacteria such as yeast cells. The Radiolaria featured on the paperback cover of *On Growth and Form* is geodesic.

Radiolaria, like diatoms, have a surrounding skeleton of silica, the mineral that figures in opals and sand, looking like glass under a microscope. Among the many drawings that Thompson surveyed was one species named *hexagona*. Since hexagons cannot enclose a sphere,

Thompson looked closely and found a few pentagons, even squares. So much for the claim that there are no right angles in living things. Radiolaria found many solutions to enclosing a sphere; there are at least 4,700 variations among their forms.

Thompson compared the construction of Radiolaria to that of snowflakes, also known as ice crystals. Symmetry among crystals occurs when two faces on the opposite side are equal, and the axis of crystal symmetry can be twofold, threefold, fourfold, or sixfold—but not five-fold. Pentagonal symmetry, so common in plants and animals (such as daisies and sand dollars) does not occur in crystals. The pattern of crystals is a result of the regular arrangement of their atoms. The same is true for snowflakes, as mentioned, based on the molecular structure of water.

Crystals, Thompson qualified, lie outside the province of *On Growth and Form*, but they do have form, and they do grow. As mentioned, an ice crystal begins as a regular hexagon but builds on each outer edge as it descends. Thompson thought the same process was true for Radiolaria and described the patterns that surround the single cell as consolidated froth, no different from soap bubbles that meet. The edges collect inorganic material as the creature moves in sea water.

Since Thompson had included engineering structures like bridges in his book, why would he note that crystals are outside the province of his survey? Perhaps to underscore his point that "forms mathematically akin may belong to organisms biologically remote, and . . . mere formal likeness may be a fallacious guide to evolution in time and to relationship by descent and heredity." The common pattern of hexagons found in crystals certainly would not relate them to Radiolaria, or to any other living forms. Lines of descent had "preoccupied the minds of two or three generations of naturalists." In the study of human evolution, the search for a missing link dominated intepretations of bones, rarely studied for the way they might have functioned until the sixties. Controversy swirled around the branches of human evolution, best drawn with a pencil with a big eraser.

The first person to use a tree as a chart for evolution was the same man who provided Thompson with the detailed drawings of Radiolaria.

Ernest Haeckel was another Wonder of Jena, where he was a professor of zoology and the most powerful advocate of Darwin's theory in Germany. In 1866, Haeckel published an extraordinary graphic of the tree of life. Darwin had used simple branch charts, but Haeckel featured branches that gnarled like a great oak, put bark on the trunk, and gave it a countenance that would have intrigued Claus Mattheck.

In perfect affrontery to fact, there are never any roots for trees that pretend to chart the roots of life. Life emerges from the base of the trunk. The general impression is that of primitive, early things at the bottom, with life becoming more complex as the branches became thicker. It is a misleading impression that Stephen Jay Gould has tried to undo for a good decade. In the eighties, Gould seized upon the image of branching bushes, but in some cases, a conifer would have been more appropriate—thick at the base and thin and pruned at the apex. There is only one branch of the human species that survived, for example, while several species have been found among our ancestors, and the branches become more numerous further back in time. The same is true for the family of sea urchins.

About 470 million years ago, during the lower Paleozoic era, there were at least twenty classes of sea urchins. Rather than the poinsettia pattern of five, some had only one food groove; it coiled around the structure in a spiral. Some forms had such a well-defined bilateral symmetry that they have been seen as ancestral to fishes, or anything with a backbone, which would include us.

The ancestral sea urchins were sessile creatures, attached to the ocean floor like kelp, immobile as a plant. All of the blueprints for modern life emerged from such ocean bottoms, and as recently as 1968, new evidence was discovered that could be added near the trunk of Haeckel's tree. As a group, these life forms are named after a place in southern Australia called Ediacara. This horizon is the province of one of Dolf Seilacher's most original ideas. He thinks these forms were an evolutionary experiment that failed. If true, long before sea urchins came on the scene, nature went back to the drawing board.

7. The Drawing Board

*If Ediacara had won the replay, then I doubt animal life would
ever have gained much complexity, or attained anything close to
self-consciousness.*

— STEPHEN JAY GOULD, *Wonderful Life*

THE thing about geologists is that they know where the bones are
buried. It was a geologist, after all, who gave Darwin the context
for his theory of natural selection. Among the treasured books that
young Darwin carried on board the *Beagle* was the first volume of
Sir Charles Lyell's *Principles of Geology*, published in three volumes
between 1830 and 1833. Lyell honored time as no geologist had before,
explaining features on the present landscape by changes that occurred
over epochs and eras and periods vast. He reconstructed a history of the
earth that was considerably longer than anyone had surmised, long
enough to encompass the curious transitions that Darwin suggested, and
a time, unimaginable and abysmal, when our ancestors were spineless
and there were no bones to fossilize. This is Seilacher's terrain, when
multicellular things began their own experiments in Form Finding.

Dolf Seilacher has been described as "the world's greatest observer,"
and he would have to be among them, since he studies things not only
long dead but long disappeared, what one of his students called ghost
fossils. These are traces left in the sand 650 million years ago, fossilized
burrows and tracks, impressions and outlines of bodies, often discontin-
uous. Imagine Neandertals made snow angels and then an ice age came

along and preserved the pattern in the snow. Imagine sediments of the Colorado petrifying the track of a sidewinder, a baffling pattern even when witnessed. Now, imagine that no human had ever seen these creatures before. What you or I might consider ghostly in the sense of poor reception, wavering and unfixed, struck Seilacher as diagnostic.

In the Flinders Range in southern Australia, the uniform size of the sand grains and tiers of ripples suggested a former beach. The area was not supposed to be fossiliferous. Fossils are usually formed in thick, rich sediments, smothered in a dense matrix of mud or volcanic ashes. Alluvial pans are ideal. Sand grains alone had never been known to create the prerequisite ooze. It was in the 1940s that geologists first noticed odd impressions on the sandstone at Ediacara, not quite as red as Ayers Rock, an inselberg of sandstone, but about the same age. What they saw were forms that look liked flower blossoms. Yet flowering blossoms appeared much later in the fossil record. Could these be impressions of sand dollars or jellyfish, washed up on the beach during a storm, and somehow fossilized by another storm?

During the sixties and seventies, several teams of paleontologists studied the Ediacara and compared these flower blossoms to the known, or index fossils, from the Burgess shale. They came up with sixteen different species of jellyfish, including medusae like the Portuguese man-of-war. They also saw things that resembled fern leaves, and reckoned they were ancient sea pens. Their impression was one of a long feather with a little disc at the tip of the quill, perhaps a root, to suggest they were attached to the ocean floor and had been uprooted during a storm. There were other things, including segmented worms.

In a bold speculation based on these details, Seilacher reconstructed the way these early impressionists grew, their parts, their manner of gathering food, their flip and flop and slide even after buried, when the earth moved them, when storms removed them. Fittingly, he ventured his interpretation at the 1983 meeting of the Geological Society of America. What he described is known as event stratigraphy, which studies many lines that have been drawn in the sand and preserved. Some things he could tell with certainty. He could distinguish, for example, not only one kind of creature from another, but the top side from the bottom, a feat since both sides were identical, and he described one creature with the construction design of an air mattress and others as doughnuts. I use the term "creature" loosely since there is some question as to whether these were plants or animals.

There were animals with long stems during the Precambrian, their heads fronded like flower cups. Other forms looked like the interior of shiitake mushroom caps, reversed and resting on the sea floor. They had no gut, so Seilacher reckoned they absorbed minerals and foodstuff from the bottom through their skin. Some even resembled the shape of sand dollars, but this was long before sea urchins, which have a rich fossil record because there was a skeleton, or shell, to preserve.

Many scientists continue to believe the Ediacara forms were ancestral to jellyfish, sea pens that resemble quills at attention on the ocean floor, and soft corals, like sea fans. But Seilacher found more than a few aberrations of this idea. Soft corals and sea pens have branches like a tree, or separate leaves, like the pinnate form of a fern. This allows them to gather oxygen and nutrients, the same way individual cells of a sponge gain access. Seilacher said the fern "leaf" was actually connected segments, ribbed, with no spaces between the compound leaf, and was the first to describe these trace fossils this way. He also noticed ribs and the suggestion of a connecting skin on the segmented worms but found them in dimensions that didn't suggest worms at all. They were not cylindrical; they were like pancakes. Other impressions of "jellyfish" no longer qualified. Jellyfish move with an undulating bell and feed by contracting and pulsating a muscle on the rim, with radial grooves toward the center. Seilacher found the radial grooves on the outside of the rim. His insights threw a hammer into the previous classification, since the structures, as he saw them, denied the manner in which their descendants ate, breathed, and functioned.

One of the fat, quilted forms from the Ediacara horizon was named *Dickinsonia*, perhaps to mark the best of times, the worst of times. In Seilacher's view it got worse, since he thinks most of these things represent whole branches of early life that have no descendants. This remains an unproved theory, but no one disputes that he noticed things that others overlooked. It's as if you took him to Umberto's Clam House in

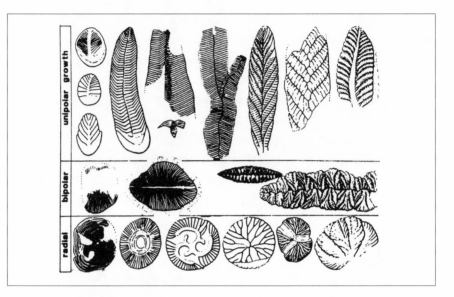

These are Dolf Seilacher's interpretations of forms reconstructed from trace fossils in the Ediacara horizon. He suggests that some may have had little disc attachments to the ocean floor.

Manhattan's Little Italy and he could tell you by which door the hit men entered, even though the bullet holes have been filled and painted over. Like D'Arcy Thompson, he could relate the remaining crater to the passing burst of a bullet. It would be a tough job, however, to distract him from the clams on display at Umberto's.

He is fascinated by the ribs and patterns of mollusks and generated his own patterns in an experiment that he called "Morphology by Candlelight." By allowing wax to drip into a bowl of cold water, he recreated the basic configuration of living shells, replete with the curve of log spirals that shape the outer rim. The British journal *Nature* rejected his paper on the subject, but it has since been published. The models were so lifelike that when he popped open his cigar box full of waxy configurations, his audience in Stuttgart emitted a gasp. It was a hit among these architects and paleontologists searching for models, but, as he confessed, an accidental discovery made by the mother of one of his students, who set the candles in cool water for safety. Yet the similarities were obvious to everyone and confirmed an axiom of geologists: I wouldn't have seen it if I hadn't believed it. Seilacher believes they hold a clue to the formation of skeletons—the big change that occurred in the Cambrian, the introduction of bones to accompany soft parts. When the wax hit the water to cool and firm, he saw the form organize itself, the same way hexagonal plates of calcium organize the plastic test of the sand dollar.

Seilacher avoids superpowered microscopes such as the scanning electron microscope, and does not pursue analyses of isotopes or chemical probes. He insists that the best tool is the human eye. For this, and the busy cerebral mass behind his orbits, he won the highest international prize a geologist can receive. He said the award is "not a personal thing; it is for a field that was very much in the backwaters." And while some of the $360,000 award will be applied to personal use ("I could use a new watch, after thirty-five years"), the bulk of the award will go into further research, beginning in Australia, then Namibia, then a site high in the Andes. At the age of 67, Seilacher has plotted an itinerary that would daunt the youthful. Equally amazing, he immediately invited along a scientist with an opposing view of the Ediacara as his guest.

In May of 1992 I took the train up to New Haven to attend a symposium at Yale's Peabody Museum of Natural History. The museum has been Seilacher's "intellectual center" for nearly a decade, where he recharges between field investigations in localities that have included India, El Salvador, Northern Iraq, Iran, Libya, Lebanon, Japan, Panama, Malaysia, Chile, the Philippines, Egypt, Australia, and Namibia. In North America, his fieldwork has taken him to Newfoundland, the Grand Canyon, Wyoming, Colorado, and Arkansas before anyone but geologists knew where the state was. Since 1987 he has divided his academic work between Yale and the University of Tübingen near Stuttgart, where we first met at the SFB conference. Prior to this, his reputation surfaced in Nairobi in 1986, when one of his students, Martin Fischer, praised his interdisciplinary approach. Dolf Seilacher's expertise may be fossils, but the matrix for his genius is geology. Fittingly, his office at the Peabody Museum is connected by a series of labyrinthine halls to the nearby Kline Geology Lab. Even his fossil casts look like pieces ofthe earth, because that was all that was left. The organisms of the Ediacara desiccated like the internal body of a sand dollar, mostly water.

At Yale, I was informed, affably enough, I would need to return again the next day if I wanted longer interview sessions, and the earlier the better. "I have suddenly so many interviews," Dolf said, in an uncharacteristic muffle, "to do with this Nobel thing." The Nobel thing turned out to be the prestigious Crafoord prize. Previous recipients include Claude Jean Allegre and Gerald Joseph Wasserberg, who shared the award in 1986 for their work in isotope geology; in 1982, the prize was split between oceanographer Henry Stommel and Edward Lorenz, a leader in chaos theory. The Crafoord is given by the Swedish Academy of Sciences to honor research in areas not covered by the Nobel: mathematics, earth sciences, astronomy—the latter "presumably what Dolf is getting it for," deadpanned Yale professor Leo Buss, taking a light-hearted stance in red hightop sneakers.

Dolf Seilacher is in the same mode. He wears an Escher tie, white cotton socks that feature colorful dinosaurs and crocodiles, and apologizes that his last haircut was in Namibia. His regulation flat top, reminiscent of H.R. Haldemann, has grown only slightly uneven since his expedition in the deserts of this newly independent nation, formerly Southwest Africa, where fossil bones from the Miocene are often exposed by blasts from diamond miners. There, in much older sandstone and deep sea exposures, he found more Ediacara fossils, making casts in the company of ostrich, gazelle, and baboon. It is his facility for such travel, and learning languages—being a student himself—that is part of his appeal to students.

He joked about his early instruction, from a Major Brady at Flagstaff, who "tried to teach me English, with what success you know! He said you must be careful to fit the right word to the situation. There was a woman who came into the kitchen and found her husband kissing the maid; she exclaimed, 'Richard, I'm surprised!' and he said, 'No, darling, we are surprised; you are astonished.'" To Yale colleagues, this was Seilacher's segue into his reaction to his award: "I was astonished . . . and it's all the more surprising that paleontology got it, and even more that a person who sometimes confesses to be a computer virgin—I leave that to my wife." Edith Seilacher was trained as a paleontologist, and in addition to entering her husband's data into a computer and processing his papers, she offers constructive criticism. Like most spouses who work behind the scenes, she was an absent partner to the prize, carefully noted in a toast by one of his Yale colleagues, but would attend the award ceremony in Stockholm.

When he finds a trace fossil in the field, Seilacher's first step is to draw what he sees. Upstairs in his office at the Peabody Museum, he reaches for a battered leather case that looks as if it might hold a clarinet or another disassembled cylinder. "At every airport I always have to open my suitcase for security guards; they think it's some strange thing." He unfolds what looks like a tripod with only two legs, a giant drawing compass of a stainless steel. It is a *camera lucida;* he uses it to measure and then reproduce the details of a form. "It is from the end of the last century," Seilacher says proudly. "I always have it with me. I put it at any table, and I put my fossil there, then I draw it. And by drawing, I observe." Drawing forces one to repeat every line and contour, to think about the thickness, to notice flaws in symmetry. Once, when I asked Mary Leakey about recognizing tools discovered at archeological sites in Africa, she replied, "Well, I've been drawing them since I was eleven." By drawing stone tools, she would notice how they had been flaked in patterns that distinguish the various tool cultures. It was a basic way to think about how things were made.

"This kind of approach is frowned upon," Seilacher said. "It's the old-fashioned approach; it's not computer; it's not electronic and it's not microscope. I do not . . . of course I am happy to have a microscope, but I say the eye is the most essential tool that we have. I do not say this is the only way." His papers on sand dollars feature microscopic views of incredible detail, so he doesn't ignore the value of that kind of data. But details don't add up to concept. He folds away his *camera lucida,* and in light of his sudden fame allows "perhaps they will come back on the market."

As a schoolboy he was interested in dinosaurs but encountered an amateur paleontologist interested in "dirt paleontology." It was from this paleontologist that he inherited his quaint instrument. "Inherited!" he corrects himself, "I bought it from his heirs!" Trace fossils were not considered "useful," nor the focus of a viable career. "At that time there was little reference in textbooks, there were no index fossils. They were considered miscellanea. That appealed to me."

To this day, few scientists study trace fossils, and Dolf entered a dark alley when he chose it as the subject of his dissertation. "Fortunately I found a supervisor who said, 'I don't know a thing about this. It is way out of my field, but I would like to supervise this study.' So I left my studies in vertebrate fossils and dinosaurs. But then I could not understand anything without delving the process."

When Dolf Seilacher delves the process, he becomes the creature. At the SFB conference, he alluded to the hopes and dreams of barnacles. If he were a barnacle and saw a whale breaching, he would want to "fly" on that whale, to expand his feeding range far beyond a stationary attachment, to avoid the corrosion that occurs on the hulls of ships, to escape from "predatory grazers." He then described the unique attachment of whale barnacles at length, but now, months later, I returned to the prologue to make certain I had heard him correctly. (Journalist's axiom: Just because you have it on tape doesn't mean that's what they meant.) So, do you think, if I were a barnacle, how would I solve this problem? "Of course," he relished. "People say this is not scientific, but it is the only way that we can best understand. One student in England said to me, 'It's strange, you always want to identify yourself with the organism.' Another comment on my approach is that I was once told I think like a schoolboy." He rightly took this as a compliment, and later, imitated a hamster on a treadmill to demonstrate a point. If he could simulate the earth's tectonic behavior, he would.

Most of us think of tectonics as the movement of the earth's plates, a gargantuan and slow process. Seilacher uses the term on a smaller scale, to describe the movement of sands and muds that occur on the shoreline during a storm. He considers these buried events as behavior and not the least bit linear. "If sedimentation were a linear process there would be no stratigraphy! It would all just accumulate, without clear phenomena. It would be indiscrete, without clear beds, without entities in the record." The business of stratigraphy was guided by Steno's law from the 1600s, that the earth accumulated sediments like a stack of newspapers, the most recent on top. But millions of years of erosion and deposition present a migraine-sized puzzle. Nowhere is this earth a land in repose. With uplifts, former ocean beds are exposed on lofty plateaus. The Burgess shale in the Canadian Rockies positions sea creatures 8,000 feet above sea level. A bed of ostracods or diatoms can denote a former lake in an arid basin. By identifying the different beds, geologists can reconstruct the history of the earth.

"The whole thing of event stratification is not whether I can identify this kind of deposit or that kind of deposit." Instead, he relates the behavior of the earth to the local biology, like the bullet hole to the bullet. "Say you have a community of bivalves, and the first storm comes, takes away the mud, leaves a pavement, or leaves a layer of shells. This may have consequences on biology, because on a shelly bottom, now the next generation can be very different." With a layer of shells, the new community would be less likely to be buried in ooze, leaving a fossil gap. "So there is no record of the last community." He can tell from the patterns what sort of disturbance affected the community. A storm has great turbulence from crashing waves, but a flood deposit is shaped by currents.

"Even when we can't find the fossils, we get information by studying the behavior of the sediments, like the the deposition that occurred during the end of a storm. First, there is erosion, the energy rises, then after its peak, there is deposition. A storm by definition starts with erosion, but what we sometimes find is one sand layer with the ripples on top, and the next storm casts the ripples! So the ripples were sort of fossilized. How can you erode sand and not turn it into grains? The grains became a conglomerate, like beer mats [coasters]. Or you find sand that is bent like the sole of a shoe? That it was bendable means it was held together by microbial glue."

"Here you have a ripple surface." He shows me a cast that looks like a tier interrupted. "Now comes tectonics, with pressure of some duration. Here is one ripple valley fading out, one fading out, one fading out, then you have a rupture, there shearing. This," he lays his hand vertically across the cast, "was the direction of stress. Here comes the stress, and now these bodies have new entities." One of the patterns of a trace fossil has been deformed by the motion. It was washed up on the shore, buried and fossilized, then the plate it was on was shifted by a second event.

"Tectonics creates things that you hardly recognize. That's evolution in sedimentology."

"Trace fossils," he concludes, "are merely the expression of a process." But to recognize any trace of living things he had to know the basic blueprints of marine creatures. So he worked on sponges, corals, ammonites, bivalves, sea urchins—especially the irregular, trekking along beaches in Malaysia and Panama where sand dollars mock the rush hour, tailgating, bumper to bumper, moving past oncoming traffic at a few decimeters an hour. He also studied their evolutionary history, looking at fossils from the Jurassic, when sand dollars emerged and the five pyramids of Aristotle's lantern became flatter and enlarged, to form a pentagon. Dolf Seilacher is not alone in thinking the sea urchin is a "mineralized pneu"; in 1986, Jacob Dafni wrote the "entire test behaves like a balloon." Seilacher wanted to know how that balloon changed into the slender form and led to "buttresses" that support the dome. Maybe buttresses are a misnomer in this case, too, but tensile support like the trunks of banyans. This kind of study was the basis for Seilacher's work in Construction Morphology, which he now prefers to call Morphodynamics. Dynamics, after all, is the terrain of physics. Hydrodynamics, aerodynamics, thermodynamics—nothing is static. Seilacher added morphodynamics. "I was led back to this process-related view, because it was all the expression of growth processes. Adaptation and growth fit together. So perhaps it's no accident that I started with trace fossils, which are fossilized action."

Seilacher describes his approach as observing and modeling, observing and modeling. "Modeling not in terms of names, but in terms of processes. Model building is essential, even the words you use to describe things have models. For everything else, you pull in global models. You see something, and either you say I don't want to understand it, or you seek shapes and functions that tell you more about it." His "Morphology by Candlelight" models repeat a process of growing by increments, even though they were not internally inspired and produced only an exterior. Under the influence of D'Arcy Thompson, Seilacher was looking at the physical forces and the generating pattern, trying to find a fundamental cause. Few things are more fundamental in multicellular things than hard and soft parts, or a balance of tension and compression, and this is what he found.

"Of course we must look for the most general rules possible. It's much safer not to, to address only the specific and describe it, without resolution. Therefore many people don't care about the unifying theories, and maybe it's too enthusiastic now about the possibility of a general dynamic theory. But pure description is just a still photograph of evolution. You look at a shell, you measure all the distances, you count the ribs. And you can't be wrong, but you don't understand it. But if you ask, Why is it curved? How does it function? you make models to test function.

"With every fossil we should ask, What happened? How did it grow? How did it live? What has happened in evolution and development? What has happened in the environment? Was it deep sea? Was it a river? Every fossil is an accident, but after death it's not over, because the remains go on changing and changing again. So we must be paleo-detectives."

He calls the science of taphonomy "life after death." Taphonomy is from the Greek *taphos*, meaning burial. "For us, organisms are not just once-living things. Biology is introduced into the geological record, then geology works with it. You have certain material, and physical forces work on that material. The detail of the organisms, what happens to the soft parts, what happens to the long bones—long bones of vertebrates behave differently from vertebrae. One can roll, the others cannot."

Teeth that roll in a river, for example, become rounded like pearls. Since they are harder than bones, teeth fossilize readily. Skulls are often smashed to bits by the compression of sediments. In rich volcanic sediments, flesh can be preserved. No bones existed in the 650-million-year-old horizon that Seilacher explored. So he relied on the sediments themselves for clues. "Paleontology should not be separated from geology, because to do so we would lose all these dimensions. I was always interested in both." And he happily fuses them, as does the Swedish Academy, which lumps them together under earth sciences.

"He's wonderful to work with—in the field, in the lab, writing," says graduate student Friedrich Pflueger, who comes from the same region and near Dolf's hometown of Gaildorf, population about 2,000 ("and three castles," Seilacher later adds.) Pflueger migrated to Yale to follow his teacher of six years. "No other students have followed him here yet. He's not been too popular in the last year because he is often gone, and some students want more supervision. His science is not mainstream at the moment. Most students here do mainstream geology, and for those staying in university [academia], they would find themselves outside the good boy network." Pflueger adds a big "but." "With the kind of kick he gives you, you go further. He is competent in every field. He's terribly good at extracting information. He can strip things down to the essential. He dislikes 'beautiful' and says, 'Why?'

"He's cheerful, and there is always a lot of personal touch. At his home in Tübingen, there are always students as guests, so different from here, where you never meet the family, never go to the home. He does not drive cars, so we are always taking him somewhere. He has a motorbike in Tübingen. So if you want to travel home with him you take the back seat on the scooter. His wife drives." The Seilachers have a daughter, Uirike, 34, and a son, Peter, 20. "Their home, their family; this is what makes him what he is. The influence of his wife is probably much greater than you would believe, much greater than he would admit, because she is—without her support all these years—she's important in discussions. She's very critical to anything he writes."

"The tragedy is," Dolf Seilacher reflected, "I married a paleontologist, with the foreseeable effect that she did not want to become the unpaid field assistant to her husband. So her drive for independence drove her away from the field." Edith Seilacher worked in clinic planning. "Only now, after she is retired, she helps me in the field." Later, when we were discussing the bias of scientists based on personal histories, Seilacher said, "You must realize that every scientist will have a biased view, biased by his own stature. I include myself or anyone. Of course you want to set yourself apart from somebody else. Because scientists are, well—have to be in a way—egocentric. So the reactions may sometimes be not only a difference of convictions but may be what we call in biology/ecology character displacement. As I described my wife, that was a case of character displacement, that she drifted away from the field." His own sense of self was displaced during the Second World War, and he became grave when speaking of his experience.

"I was eighteen, nineteen, twenty, having grown up in a regime in which your opinion was so inoculated you just didn't get any other opinion. You had no foreign radio, no foreign newspapers. A system that wants to do it can manipulate a person against reason. You had only this view that was meant to reach you. Therefore it was the traumatic experience of the years before and during the war. So we fought, and there was the complete collapse of this artificial system—the pseudo-skeleton and the nucleus went—everything was floating. Fortunately, I was never in a place where it became human beings versus human beings; it was always machines against machines, but implicitly people behind [the machines] were involved. I was in the Navy; I was a wireless operator, so I didn't have to shoot. At least I knew I was already interested in paleontology, even as a boy. And I knew there I had firm ground."

He cultivates Pflueger's confidence, supervising his thesis on the firm ground of sedimentology, a broad field that includes deposition from glaciers and deep-water silts as well as the denuding of rocks. "I would not supervise a thesis in my field of specialty, for egotistic reasons. I can't learn as much if it's in my field. I learn a lot discussing things out of my field of vision, so I broaden my view. In the interest of the student, he has a chance, in a few weeks, to be ahead of this old professor experience-wise. So supervision is not telling him exactly what to do but more independent. The self-confidence of a student can grow better out of the shadow of the master."

Over lunch at a local Chinese restaurant, when Seilacher repeats his disdain for statistics and electron microscopes, Pflueger offers him a fortune cookie and says, "Although, I guess we are not ruling out modern techniques as a principle." To which Seilacher booms in response, "Of course not!" He then reads: "You have a deep appreciation of the arts and music." "You are about to begin a long journey," would have been providence itself. After lunch, Pflueger told me, "You know, he has received another award," this one from Japan, to lecture in Kyoto and at the marine station in Okinawa for three months. "It's as if he is harvesting what he has learned all these years."

Showing me some casts in the lab attached to Seilacher's office, Pflueger recalled: "Once, we were looking at these sediments, with trace fossils, on a cliff, and Dolf said, Oh, that's nice. So we know the current came from this direction, what kind of environment, that it was buried for some time. He made it all so obvious. A colleague with us said, Shouldn't we measure this, and shouldn't we compare this data statistically, do the standard deviation. He was in love with the computer, and reconfiguring the data. Dolf doesn't smile. This drives Dolf crazy. It's like giving something a name, and then because you've given it a name you can forget about it. This is what I've been learning from Dolf, to ask why, to understand it."

Wham! Bang! Professor Seilacher enters the room with a broad smile, slams the door behind him, and completes the thought, "It's almost like saying beautiful." Pflueger chimes in, "Beautiful, yes. But why?"

As I follow Seilacher into his office, he announces, "I have to close the door, because after lunch I am allowed—well, I am not allowed, but I will have a cigar." He opens the window ("totally against regulations, of course!") lights up a Milde Corona, then offers me one. I ask him about his upcoming stint in Japan. Beyond his lectures, he says, "I don't have a fixed program. I'm just open for new experiences. It's human interest in Japanese culture which is so fascinating, because their way of not giving up their old home culture. Be Western in the office, go home and take a hot bath, be Japanese."

Confirming part of his fortune, he says, "For instance, when you see a Japanese watercolor, concentrating a whole image in a few lines. This would be a counter-experience to the dinosaurs in dioramas, with too much put into them. Imagine," he leans forward and draws a parabolic line with his cigar, "a Japanese artist making a dinosaur, conveying this thing. I think more could be done in that way. Creativity is neglected by many scientists because it's not productive, it doesn't add to the CV or the publications list, in this very competitive world." Like Einstein, Seilacher believes in the value of imagination in forming theories, and when fishing for a word for evolution, he seizes upon "creativity," with a dynamic nuance that, Major Brady would have noted, distinguishes it from "creation."

He shows me new casts from Namibia. "We make casts and leave the fossils where we found them, removing no national treasures. If you remove the matrix you destroy so much information, how it was positioned in the sediments. This is from inside the bed," Seilacher continues. "These are not typical of those found at Ediacara, in Australia. You see here a deformation. If the organism was hard, it wouldn't bend. It would break! Yet these fossils maintained their original orientation and are three-dimensional. I think this species lived like a root system within the sand. If you compare this to anything today—like sea pens or sponges—that kind of organization does not fit their feeding behavior. The quilted chitin allowed the form to be expandable, with a net frame that maintained their shape and maximized their surface area for life-sustaining interaction with the environment. To feed, they were absorbing nutrients from the water—or the ones with the stem, standing up, presumably using light."

Seilacher's horizon is "a time of the earliest life on earth, when animals really arose." Of the more famous horizon known as the Burgess shale, he says, "No, no. That's too young." Some fossils from the shale in the Canadian Rockies reveal the first hint of a backbone.

At the symposium the previous day, Stephen Jay Gould described that "young" era, a mere 530 million years ago. "The Cambrian was that unique time, never to be found again, in which the ecological barrel was built, and there was an untold possibility for the expression of possibilities. Everything happened and the experimentation would sort it out later. . . . It was never an empty world again, and it was into that empty world that this vast set of possibilities radiated." Of these possibilities, Gould said, "Designs locked later, but they weren't locked in the Cambrian."

Gould likes to "replay the tape" of evolution and speculate on what might have happened downstream if some key events had been reversed. What if the designs of the Ediacara had locked? We might not be here, he reckons, one of the reasons why he refers to our presence as "a glorious accident." Here's why.

The quilted and pancake forms that Seilacher described struck him as "an alternative solution" to a design problem of gaining enough surface area as size increases. Gould relies on the physical laws of growth that

D'Arcy Thompson emphasized. Volume increases to the third power, while surface area increases only to the second. Seilacher reckons that the air mattresses and *Dickinsonia* of the Ediacara fed through their surface. To grow larger, they would have needed something to support this blob, a function of the ribs and quilted pattern, and the skin that Seilacher thinks was like chitin, the paper-thin but remarkably strong coat of insects. The net worked like the string around a soft cheese. Design alternatives include the internal pressure of a pneu, or an internal skeleton. The largest living thing on earth is neither a giant sequoia nor a blue whale, but a fungus, *Armillaria bulbosa*. It occupies 15 hectares, weighs 10,000 kilograms, and may be 1,500 years old. It's not going anywhere, except on the edge of its margins, where it grows, supported by the earth itself.

Our heart and lungs have the fractal pattern that achieves decompounded coverage of great surface areas, and even the sand dollar has a little mouth and an internal hydrology system for processing its food. The emergence of the backbone, and the development of internal organs (such as gills and lungs) was a creative solution to a design problem that resulted in all vertebrates. Once bones were formed for internal support, they became useful for other things, like locomotion, a creative use of available materials. But if *Dickinsonia* became a blueprint for life, would it foreclose the onset of bones?

Gould displayed a triangle chart that Seilacher devised, to show the forces of creativity.

SEILACHER'S TRIANGLE

historical contingency of phylogeny

functional   *[constraints]*

active adaptation   *formal rules of structure*

This is an attempt to categorize the different influences that work to shape any living blueprint, ancient or recent. A tree would have a long bit of data at the apex of the triangle, because it has a long history and has diversified from its original blueprint, growing taller, for example, than any other plants. Its lineage can be traced to algae, a single-celled plant. On the left side of the triangle, functional, active adaptation would include photosynthesis, a habit inherited from algae. Constraints include the fact that a tree cannot grow above a certain height; it cannot walk or see, and the shapes of its seed and leaf are more or less dictated by its ancestors.

(Seilacher has since refined this triangle by adding a third dimension, environment, which would include such factors as a trunk that grows stronger under the influence of winds, special adaptations for drought, and response to cool temperatures among the deciduous. It was an important addition based on his studies of trace fossils, because he found that many designs were influenced by their habitat in the ocean.)

"For nongeologists," Gould continued, "let me tell you why Dolf, who, like me, has a geological background, plotted this in a triangular diagram. We're so used to plotting rock compositions in this kind of matrix. The whole point is that most things are in the middle, not that you should believe that these end members have independent status. You're plotting actualities." He pointed to the right corner of constraints. "What do you call a constraint such as DNA? In one sense, it's as universal as can be, because all life is made that way. In another sense, [he pointed to the top of the triangle] it's historically contingent on life being structured of DNA and not in another conceivable way. That's what I mean by saying that what we consider interesting falls not on the points outside the triangle, but in the domain in between. Constraint is exerted not only by the immediacy of one's own phylectic heritage, but also the structure, the architecture of life." If the architecture of our ancestors had been the blueprint for *Dickinsonia*, would we be here? Gould suggests not.

Seilacher continues the theme in his office. "Basically it's so obvious that there are constraints. We cannot do everything. We cannot grow hair if the material is not available. The barnacles I described in Stuttgart cannot attach to everything. One kind of barnacles I did not mention live not on whales, but manatees. But the situation is the same; it's mammalian skin. You learn that any member of the seal family was never inhabited by barnacles; with the hair they cannot get attached. You need a naked mammal like a whale or a manatee." So constraints have to do not only with the architecture of the creature, but its environment as well.

In North Carolina, Seilacher found trace fossils from greater depths. "These little trails were made by small wormlike organisms," he says. The trails zigzag, circle in a coil, or trace an outline like fingers of a glove. "I then identified myself with the organism." So, if you were a grub, and wanted to consume the greatest amount of food in the shortest distance, you would not waste energy with random excursions, but move the way you do when you mow the lawn. Seilacher describes it as "systematic strip mining." This, he says, "indicates a behavior program, this means a very high neurosystem."

David Raup, of the University of Rochester, played with the parameters of these patterns with a digital computer. Seilacher supplied the casts. The computer worm was granted capability of four movements. It could move straight ahead, make a 180 degree turn, or turn toward or away from an existing track. Raup got a variation in patterns that he and Seilacher think are "comparable to genetically controlled" behavior of different creatures that made the real tracks.

Seilacher proudly displays a drawing inscribed "To Dolf From Dave," and repeats, "I am a computer virgin." The conclusion in their joint paper for *Science* partly explains why: "The primary value of the simulation studies is not to be found on the level of factual results. The mere presence of the computer method encourages rigorous analysis of meander and other patterns in trace fossils." It was Seilacher identifying with the organism that came up with the four determinations; and after some experiments, the results matched the real trace fossils.

Seilacher's rigorous study led him to suggest that unrelated species tend to respond in similar ways to similar habitats. The depth of the water is an obvious influence, both physical and environmental. Worms at shallow levels make more irregular trails, or simple dwelling tubes,

The real worm burrow patterns on the left were generated by computer in an exercise by Dolf Seilacher and David Raup.

vertical and U-shaped. Nutrients on shallow bottoms have patchy distribution. The burrows are more refined and developed in deeper waters, where the distribution of food particles is rich and relatively even.

Similar kinds of sea worms survive today, as do other forms like those in the Ediacara horizons. Sea lilies are attached to the ocean bottom with stems, and for a while the ancestors of sea urchins were sessile, with no means to move. What are the design solutions to this kind of anchorage? Mimi Koehl investigated forms influenced by their position in the water. In the mode of D'Arcy Thompson, she studied their form based on physical influences, including forces that create drag. Viscosity happens because water molecules resist sliding past one another when confronted with an object in their way. The result is a boundary layer creating eddying wakes: the smaller the wake, the smaller the drag.

Mimi Koehl teaches biomechanics at Berkeley, and her innovative approaches led to her selection for a MacArthur grant in 1990. She receives $52,000 a year as a MacArthur fellow, the amount of the award based on the recipient's age. (Koehl was 41 when she was selected; older recipients are awarded a larger annual stipend.) Since there are no strings attached to the grant, Koehl wanted to "use it to buy time." It allowed her to "try some crazy new projects that you have to prove are possible before you can attract regular funding." One of her previous projects included testing models of flying frogs, another, the wing span of early insects, based on models made of epoxy resin and paper.

When she spoke at the SEB conference in England in April of 1992, Mimi Koehl wore a model around her neck. It was a long red scarf. She would use it to demonstrate the movement of seaweed, or kelp, in water, and how these plants manage to stay attached in the great turbulence that occurs on the West Coast north of San Francisco.

With Stephen Wainwright of Duke, she studied the giant bull kelp known as *Nereocystis*, which forms "extensive underwater forests" along the West Coast of North America. The canopy is composed of numerous long blades that float near the surface to process sunlight, but they are anchored to the sea floor. In between the blades and the anchor is a neck, called a stipe, "a structure so flexible it can easily be tied into knots without deforming." It has to sway back and forth with every wave surge and tidal flow, and it bears the influence of drag on all the blades, like the string of a kite. The blades do their part. When exposed to rapid water currents, the kelp grows blades that resemble thin straps. Plants that encounter only slow flow grow wider blades. Of course, weather can change the influence of turbulence, but Koehl did tests to monitor the organisms' reaction to many conditions. The wider blades flutter and remain spread out when confronted with high forces of drag. The thinner straps collapse into streamlined bundles.

Koehl described kelp as a "flexible floppy" form, and included other algae, or seaweed, in the same category. Based on studies in the Caribbean, she compared these to sea fans and stiff corals. Stiff organisms "pay the price of drag," so stiff coral is found at greater depths, while sea fans were described as intermediate; their flexible skeletons sway back and forth as the waves pass over them. Sometimes the coral structure is parallel to the flow to reduce drag in turbulent waters but grows at a right angle in less turbulence, to feed with a greater surface area.

In her survey, Koehl saw no reason to distinguish plant from animal; the study was one of structure, of forms that were shaped by the dynamics of the water. Coral "are marine animals that think they're plants" in terms of form. Whether plant or animal, if the form is anchored "deeper than half the distance between the crests of the waves passing overhead, it does not 'feel' the waves."

Things anchored on the shore get hammered. Koehl also studied intertidal surge channels, where she found sea anemones that were flat as a pancake in exposed areas. They avoided drag by "hunkering down." They were normally taller ("fluffy"), but being tall in these kinds of waters would increase their drag. Even tall ones in calmer water endured greater stress than the flat, round, hunkered down in the surge channels. It was an illustration of "how profoundly the shape of an organism affects the stress" factor.

The structure of the stipe for the bull kelp is different from that of other plant tissue. A stem of grass, or a tree trunk, will be stiffer on the outside to resist bending moments. The stipe is stiffer in the center, which allows it to bend parallel to the flow, with the flexibility of a leaf avoiding drag.

Koehl used her red scarf to demonstrate this, rather fluid and graceful herself, with long straight blonde hair that moved as she swayed and leaned to express the flow of a wave, stretching the scarf at full length when describing a peak surge, "a full-tilt boogie." Professor Koehl used engineering terms to describe the plant's response ("remember, deflection is proportional to length to the third power"). What she discovered was that the flexible plant could increase its length and not increase drag, *after* it reaches a certain length, say, a meter for seaweed.

"So with an oscillating flow, if you're flexible, you're not going to suffer an increase in hydrodynamic force as you grow. In fact things may get better as you grow." At first she uses only a short part of the scarf, holding one end down as if it were an anchored root. The shorter alga, upright, gets the brunt of the force, while a longer one (she extends the scarf horizontally) does not pull at the root because the drag has actually decreased with length. "If you're stiff," she concludes, "the force [of drag] increases as you grow." At the end of her presentation, Koehl said, by way of qualification: "What I have talked about is isolated things. I haven't talked about the canopy. These things can fail during a storm, and sometimes when a single kelp breaks, it will wrap around a neighbor, which then bears the drag force of two, and you find great balls of kelp on the shore afterwards."

Dolf Seilacher studies the effects of turbulence and flow in the past tense. In one cast of Precambrian fossils, he points out, "The agents that were smothering these strange organisms were clearly storm events. Here we see one feature that is important for preservation, the various modes in which this air mattress became compacted, which tells you about the rigidity of the structure. We found rare individuals which were flipped over like a carpet. This tells you something about the rigidity also, and at the same time shows you that the lower side never looks different from the upper side."

He could tell this by the quilting pattern, which is "elementary." The smallest trace fossil found was only 4 centimeters long, the largest 40;

but the basic subdivisions remain around 20. "In the very rare poor specimens that were transported and mangled up, you see a deformation. That deformation is clearly not one of a soft skin, but of paperlike material—I would say a chitin membrane, that can become creased." Other examples were "smothered by volcanic ash fall that produced some local current." Sometimes volcanic ash is transported by a river, and builds up in alluvial sediments. "As you can see, all the stalk-forms are aligned in one direction." Seilacher keeps saying "as you can see," but I wouldn't see anything unless he pointed it out. "We know from the cross bedding, the current came in this direction." He lays his palm across the cast.

In August of 1990, Seilacher's team studied exposures at Mistaken Point on the southern tip of Newfoundland. While the largest forms resembled a fat fern leaf, the most common fossil is spindle-shaped, with growth stages from 1 to 12 inches in length. It looks like a pair of well-defined lips. The structure included about twenty segments along a central axis, the crack in the lips. These segments were established early in life, while fractal subdivisions continued as it grew, which could be told from younger, smaller forms. One cast reveals how the rock buckled, a tectonic collision that occurred long after the organism was buried.

"These spindle-shaped fossils had a certain symmetry reflected by the impression on the bedding plane. With tectonics, if the body is at a right angle to direction of stress, the release of the stress was always located at the middle line of the organism. It changed with a slide of only a few millimeters. Had the organism been aligned in the same direction of stress, then you would see all the fine partitions."

Some pinnate forms recovered in Newfoundland grew by additions to their margin, like a clam, and were similar to fossils discovered in England, where disc-shaped anchors resemble expanding doughnuts called Cyclomedusa. Similar forms have also been found in Siberia. There, another form called Tribrachidium, also round, appears to have a structure approaching complexity, so maybe the scenario on the Ediacara should not be considered hopeless in terms of creativity. Other round forms, including Tribrachidium, resemble the blueprint for the bottom of a sand dollar, with grooves radiating from the center. None of

Seilacher's interpretations have been anointed yet, "Because," he pauses for an understatement, "there has been some debate."

"These were traditionally interpreted as ancestors of later compound animals. The original view was to project phylogenies back into the Precambrian, to align them with things we have today." Because Seilacher thought they were "another branch of living things with unique constructions," he called the exhibit at the Peabody Museum "Strangest Life on Earth." In a 1985 column in *Science*, Roger Lewin wrote a story on the subject with the headline "Aliens Here on Earth." "That sounds very like . . ." Seilacher pauses. "I would not have put it that way." Writers don't always have control over headlines, and Seilacher was the first to make the E.T. connection in 1983. When he made his presentation to the Geological Society, he referred to our wonder at what extraterrestrial life would look like. The face of Stephen Spielberg's E.T. was modeled on a giant Galapagos tortoise called George. How much did the physical influences of this earth have to do with these forms? Would *Dickinsonia* be the same shape in the seas of other planets? What would forms look like on planets that were extremely warm, or without oxygen? His point was to emphasize the physical and mechanical influences on forms on earth, a theme of D'Arcy Thompson's.

Dolf emphasizes that extinction is an accident that evolution cannot foresee. While this was a time when nature went back to the drawing board, the blueprints of the Ediacara were not design failures. Just because these blueprints disappeared doesn't mean they didn't work, up to a point. There are sea potatoes that exist today. We return to the idea of evolution as creativity. Seilacher compares it to a broader view of physics, with particles, elements, and atoms that interact. Creativity, he says, "means you form patterns, you form structures at some level. The question is, When do these structures become individualized so that you get particles that have variations, particles that can interact, and then build up to a higher level of organization? Now, in organisms, it's very complicated, but this is what happens. First it's one cell, then many cells, then the cells interact and specialize. Then tissues form organs, then organs form an organism, then the organism forms part of a population."

He thinks the same organization occurs with sediments, including sand grains. "Normally, one would say no, and it was actually cited in a paper, as a typical example, that dunes or ripples, because they consist of sand grains, cannot become organized; they are composed of entities that do not become fixed. Now here I have the example, ripples in layers of sandstone. They do not behave as individual sand grains, but new material." He is not alone in thinking that sand grains organize themselves. At the Scripps Institute of Oceanography, the theory of self-organization has been applied to the formation of sand dunes.

Another example arose in the same port where I found the "Legend of the Sand Dollar" on a postcard. In Beaufort, North Carolina, in the summer of 1907, a biologist named E.V. Wilson conducted an experiment with local sponges that grow in the harbor. Sponges were also considered neither plant nor animal and were once classified as zoophytes, a hybrid name for an animal-plant. Their architecture is a commune of many cells, but each cell eats on its own, supplied by communal canals. (It was this capability that convinced Seilacher that what he saw in the Ediacara was not the ancestor to sponges.)

Dr. Wilson collected a living red sponge, cut it up into bits, put the bits into a muslin bag, put the bag into salt water, and squeezed the bag. Red cells emerged through the fabric of the bag and drifted down to the bottom of a dish. In less than an hour, there were clusters of red. The cells were regrouping. Then they built vertically, and their vertical extensions began to connect like a small tree, then a forest. Once the basic blueprint was reconstructed, there was a shift in assignments, with identical cells assuming different functions in different positions, where they recreated the canals and chambers and vent tubes of a mature structure. Within a week the whole structure was completely rebuilt.

Dr. Wilson did the same to a jellyfish, and it pulled itself back together. These accounts come from John Reader's wonderful book *The Rise of Life*, which says it all in the subtitle—*The First 3.5 Billion Years*. Sponges date back to the Precambrian, about 700 million years ago, and this framework was their solution to growing a large surface area. The success of those in Dr. Wilson's experiment is reflected in their species name, *prolifera*.

At a party in his honor at the Kline Geological Laboratory, Seilacher announced, "I would like to share this award with Bruce Runnegar, because he has a completely different background, a very sharp reasoning—and most important, he does not agree with my interpretation." Runnegar and others think the forms of the Ediacara became more complex and led to current forms like worms, sea pens, and crayfish.

"Science is not about being right or wrong," Seilacher says, "it is about understanding more, and we learn much more if we put our heads together in the field. The ideal thing would be to co-author a paper, with two interpretations, and the reader can decide. What we need to do is get away from the controversies that dominated science in the last century, when people of a different opinion decided not to talk to each other."

"This field has been held up by people sitting on their information, their specimens, and not sharing it with others," Seilacher had said earlier in his office. "That is the style of the big controversies of the last century. People identify themselves by their interpretation into camps. But I want to share the information, to use the money to get to places in the world where these things occur, so we get the best information, and we get it with different eyes. The opposition has to be folded into the research system! If I don't have to defend it," he says, "my ideas will not develop."

◆ ◆ ◆

CONSTRUCTION Morphology was one of the ideas that Seilacher helped develop. The term was coined by zoologist Hermann Weber in 1955, and Seilacher revived it in 1970. "Construction Morphology was a methodology, a diagram to order your thoughts on what you were observing," like the triangle, Seilacher said. He wanted to add what he was observing—the dynamics of an effective environment. A wing is shaped by the way air behaves and to resist gravity. The slender dome of the sand dollar was an adaptation that allowed it to hug the shoreline and burrow into the sand. So now Seilacher uses the term Morphodynamics. "The distinction is very important," he said. During an SFB meeting at Bad Homburg, "people used Construction Morphology without the dynamic view." Seilacher found their view too static, not a study of change.

He prefers Morphodynamics because it implies change in a non-linear fashion. "In a way the synonym is self-organization—how in space and time things evolved, starting with the concept that nothing in this world is constant. This is not so popular here as it is in Europe, because in America you have the tradition of a very linear, quantitative approach, which synergetics is not, and many recent phenomena are not. The important thing is not that we overthrow the old approach. The linear approach was seen as a necessary way to do things, but it's not the only way. You see, what I'm happy about, in Europe we were more holistic in our approach, and 'holistic' in the old times had the under-pinnings of Vitalism. And today we don't need Vitalism anymore, because in the synergetic world you don't need Vitalism."

Vitalism is the notion that living things are destined to become per-fect designs. The architecture of hermit crabs is perfectly eclectic. With no home of their own, they use an empty shell on the beach. Often these homes are decorated with encrusters that grow to produce a logarithmic spiral tail, or "arms" that extend like wings. "This shape is the outcome of individual reactions in a global system, based on economy," Seilacher says. A hermit crab does not look economic. The economy is based on the feeding habits of the encrusters, sponges and anemones. "First, they organize themselves by the growth of the hermit crab. Then, there are these preferred positions. One is around the edge, when the hermit crab eats, where they get some food. The good position for the moment, when the crab turns around 180°, will be bad then, because they will stick in the mud. Where are the best places to avoid this? In the axis of coiling. Therefore they grow that way. So like your hamster in the treadmill, always going up [at this point he becomes a hamster on a treadmill, digging with his hands]; at any moment it tends to grow upward, never outward. The result is a system that becomes spiral.

"So this is the hermit crab story; there are so many stories, but one must always look for the common denominator." He believes the com-mon denominator in many patterns is a self-organizing principle. For example, fingerprints are not only unique to individuals, but vary among our fingers. Seilacher compares their pattern to the growth of a zebra's stripes. These patterns are planted in their embryonic source, but they stretch when the creature grows, like a freckle expanding. The growth is similar to what occurs as the stripes on a nautilus spread, and like a zebra, the final result is a little different in each individual and markedly different among species. This common generating pattern is the same that D'Arcy Thompson used to explain the way a plant unfurls. Petals evolved from leaves, and the oldest flower resembles a magnolia.

"I was always interested in extremes," Seilacher says. "When you understand extremes, you see the limits of a *Bauplan*," or blueprint. He thinks symmetry is overrated. Yet he had described asymmetrical sea urchins in the Red Sea as "pollution morphologies." "Certainly there was a change in the effective environment, and if this pollution is persistent it would lead to a change in morphology. But hopefully, by not being permanent, it is unlikely to have real effect on evolution." But wouldn't such a dramatic change in their symmetry affect their fitness? "No, because we overemphasize symmetry. Symmetry is not a biological value. Symmetry is an outcome of pattern formation." But in this case, wouldn't their shape be essential to balance, to hydrodynamics? Seilacher balks at this; in a survey of a highly successful population of sand dollars in Panama, he found a preference for asymmetrical behavior, a "longshore orientation" for feeding in parallel lines to the beach.

"The beauty of the nautilus shell, that's again our view of what's beautiful. I hate this word 'beautiful' because this word is a shortcut. You have your books that say beautiful and that's it, end of story. If you say 'fascinating,' then that's okay. That's what I want to say with these wax shells," in Morphology by Candlelight. The wax structure begins with one drop, and the beginning point is not unlike a mollusk larva. From this single point, the wax creates a logarithmic spiral, as happens in real shells. When the wax hits the water, it confronts a surface tension; the hardening form is guided by this and the hard contour of the previous margin. "What you have with the candle wax is a symbiosis of hard and soft components." Seilacher believes this kind of symbiosis influences growth among the living, and it is an analogy that suits tree-ring growth and the growth of human bones.

"In growing skeletons you seem always to have a feedback, or synergism, between two components, hard and soft. In the candle wax shell, it means continuous alternations. There is one sheet of molten wax, and now comes the next, and the next has its own shape? No, it takes the shape of the previous one, and superimposes its shape, even if in that margin, I have a little irregularity—and we observe this in real shells as well as the wax shells—there was a little something swimming on the surface and made a little indentation, just like a bite." Crabs and fish bite mollusks. The wax model "healed with the next growth line, exactly as you find it in the living shell. So it's the mutual control between hard and soft."

"This happens with sand dollars," Seilacher begins, and reminds me of his debate with Malcolm Telford in Stuttgart. "I did not say that they *are* pneus," he expands. "They are not balloons, they are not soft. It's a real pneu only in the short phases of growth. This may be an hour within weeks. So when Telford said he measured and there is no pressure difference, that is no counterargument, because you would have to measure continuously over a long time to get the peaks. Probably they cannot grow continuously. There is no, I believe there is no continuous growth in any skeletons anywhere. I think you and I grew our bones the same way [as a sand dollar grows]—not all the time. I think it was Frei Otto who made the very interesting suggestion that maybe we grow our skeleton while we sleep. Growth is by rhythmic phases."

When asked for his card, Seilacher pulled out a tiny stack of return address stickers. "Now maybe I can have some cards printed up," he mused. As I left his office, he was searching for the telephone number for the British Museum of Natural History, explaining that he left his address book in a New York City cab. I suggested he might want to enter his address book data into a computer as a backup, and the man so keen on self-organization began, "I am a computer . . ." I finished, ". . . virgin."

8. Economy of Motion

T HE controlling metaphor intended for this book was the form of trees. Then a sand dollar volunteered for the title. But the gist that continues to interject itself is the "Vision Thing."

Dolf Seilacher emphasized the value of unsullied observation. Pythagoreans saw numbers in plants and planets. The accuracy of Leonardo da Vinci's view of the flapping wing folded in and out of favor. The Wright Brothers eyeballed turkey vultures to solve a key problem of lateral control, and Paul MacCready made some quick calculations for the Gossamer Condor after sighting a couple of soaring birds. Jeremy Rayner had to depend on photographs to prove his theory about the upstroke, but the precise image of a flapping wing was immaterial to his original thinking. Rayner does not have stereo vision. D'Arcy Thompson saw geometry in forms where others didn't. Buckminster Fuller recalled the influences of being "born cross-eyed" but sensing the strength of a triangle. An SEM photograph of a fly's wing with its three-speed gearshift proved at a glance what I spent pages trying to impart. Claus Mattheck walks through a forest with open eyes, blind bats can identify their prey, and people with migraines see geodesic forms. The latter, along with several other phenomena has been posed as an example of self-organization, but we have to wonder if it is just a trendy template, if people are seeing what they hope to find.

The organization of the eye in most animals is about the same. Rods and cones don't vary in size, but the dimensions of the retina and pupil are adjusted for the reception of light. So the proportions of an eye don't graduate by the same rules that influence other body parts. A human eye is about $\frac{1}{200}$ of the total body size, a whale's eye is about $\frac{1}{3000}$. But a mosquito's eye is about $\frac{1}{8}$ the size of its body. If the proportions and structure were the same for insects, the pupil would be too small to gain a clear image from diffraction. It was a design problem.

The solution was a compound eye composed of many units, each optically isolated. D'Arcy Thompson pointed out that because of this spherical mosaic, insects do not perceive light straight ahead, but somewhat off-center and marginal; in darkness they will head for a light at a certain angle because the beam is perceived from the side. Consequently, a moth considering a flame will approach it from a series of angles, homing in to the light source by a flight pattern that creates a log spiral, ceasing by increments.

Dolf Seilacher saw no evidence of eyes in the Ediacara, but in the Burgess shale there was a creature called *Opiabinia*, with five eyes, two paired, one centered. Some of the experiments from this era remain; crabs have eyes on stilts.

Darwin developed a fever when thinking about how the eye evolved, especially the automatic focus for changing depths of field that we encounter rapidly. Scan from these words across the room and out the window, then back. In feature cinematography, there is a special assistant to the cinematographer, a focus puller, to recreate this kind of transition. Such a practice is possible in movies, where each shot is rehearsed, and the barrel of the lens is marked with tape beforehand. Eagles and other birds of prey seem to have something akin to a telephoto lens. Blind snakes perceive heat, and the infrared images of a mouse denote the shape of its face. The simple principles of reproducing an image by a pinhole were recorded in Leonardo's notebooks. He noticed that a small window in his dark studio created inverted shadows of people passing by on the street, and wrote of the potential for capturing images in a dark chamber, or *camera* in Italian.

Now the camcorder has become a handy tool for biology professors, and during a Society of Experimental Biology (SEB) conference in England, Harry Bennet-Clark from Oxford explained how the lens from prismatic binoculars could be attached for macrophotography to avoid putting the equipment so close to the subject. For studies in biomechanics, the camera imparts details of motion and materials. Scanning electron micrographs reveal the organization of muscle fibers and details of skin and bone, the construction of protein, the bonds of tissue. Two New York physicians recently received a patent for an organic glue that can be used in place of stitches to close a wound. The adhesive, which is applied with a laser, is similar to the natural compounds found in connective tissue. Because it provides a watertight seal, it is especially useful for eye surgery. That is U.S. patent 5,209,776. Number 5,179,963 is for a tiny balloon that can be inserted to relieve repetitive stress injury that can result from continual typing on a keyboard. Instead of splints or medication, the pneu replaces the lost elasticity of the wrist ligament and keeps it from pressing on a nerve.

It is easy to perceive the elasticity of our own skin, and tendons were once used as the spring for Roman ballistas, to propel 90-pound stone missiles a quarter of a mile; in a pinch, hair was used as a substitute for tendons. More difficult to imagine is that solids such as steel are elastic. When you press your finger on steel, molecules rearrange themselves as a measure of your touch. The molecules get busy and reorient themselves, deflecting the pressure. "When you climb the tower of a cathedral it becomes shorter, as a result of your added weight, by a very tiny, tiny amount, but it really does become shorter," wrote J.E. Gordon in *The Science of New Materials, Or Why You Don't Fall Through the Floor*, his sequel to *Structures, Or Why Things Don't Fall Down*.

The British physicist Robert Hooke (1635–1702) devised ways to measure this elasticity, and both biomechanics and engineers employ Hooke's law. His words "As the extension, so the force," denoted that deflection is in proportion to the load. It is Newtonian in the sense that when you step on the floor, the floor pushes back in proportion to your weight.

Hooke got a little more Newtonian than Isaac Newton preferred, citing a letter he had written to Newton outlining his own hunch on the law of gravity, but he lost his argument for plagiarism; consequently Hooke is remembered for other things. He invented the iris diaphragm used in cameras today and put an unknown world into detail when he invented the compound microscope. It wasn't the first microscope ever invented, but he managed to combine a series of lenses, finely polished, each magnifying the one before, and the arrangement lent a new precision. Hooke coined the word "cell," because what he saw reminded him of the maze of monk cells in a monastery.

Now we can see and photograph the elastic parts that allow a flea to jump 15 inches and a locust more than 30, with muscle filaments that extend longer than ours. The key to their success is stored energy, working like a spring, which a caterpillar demonstrates with no legs at all, coiling itself up, then releasing to spring 5 inches. The spring allows energy to be saved and used again, outwitting entropy. Leonardo had a blueprint for a spring-driven car in his notebooks, drawn up long before springs were used by Swiss watchmakers. D'Arcy Thompson attempted to explain the lag: "It behoves [sic] us always to remember that in physics it has taken great men to discover simple things."

Hooke's law was refined by Young's modulus, a way to measure the elastic strength of a material. When a wooden floor pushes back, it recovers its stiffness based on the modulus of wood. All kinds of material can be rated for their ability to recover, from silk threads to the stipe of kelp, human hair, and steel.

These kind of measurements were combined with the larger view of economy for locomotion—sustainability winning over burst speed—and formulas relating fuel and oxygen consumption to distance traveled over time. Fish use both sustained and burst speeds, but the introduction of legs added other measurable factors like the influence of stride. Colin Pennycuick, who studied soaring birds in the Serengeti, also filmed African mammals. He found that the stride frequency was inversely proportional to the square root of the leg length. This means if a camel has legs nine times as long as those of a cat, you can expect it to work its legs at one-third the frequency. Yet with the naked eye, it was difficult to see the legs at work, even in something as big as a horse.

British photographer Eadweard Muybridge is sometimes credited for the invention of motion pictures, an honor granted to Thomas Edison in the U.S., and the Lumieres and Etienne-Jules Marey in France. What

Muybridge did was shoot a series of photographs in sequence by assembling a bank of twenty-four cameras to capture that many sequences. There are 24 frames per second in a movie film, the pace at which movement appears to flow naturally, yet Edison's early cine had 18 frames per second, hence the hasty pace of Charlie Chaplin. Muybridge did not achieve such speeds, but he did manage an apparatus to convey the sequential. To merge his still photographs, he devised a viewer that he called a zoopraxiscope, a revolving disk that created a moving image, albeit halting and uneven, like one of those little paper books of a batter taking a swing, found in cereal boxes and thumbed to make the pages flutter. If you could make them flutter at twenty-four pages a second, you would have the basics for Disney's animation. The equipment Muybridge invented is in the Smithsonian Institution.

The zoopraxiscope, as the name implies, revealed animals in motion, including humans. Muybridge published his photographs in a series of volumes, beginning with *Animal Locomotion* in 1887, which intrigued both the art world and the scientific. The photographs featured simple action, such as a mule kicking, a woman leaping, a cow walking, and Muybridge himself striding from stage left to right in the nude. The images were placed in a grid, suggesting continuous action. In hindsight, the black and white enhances their documentary style. During a 1992 exhibit of Muybridge's work at the International Center of Photography in midtown Manhattan, a reviewer wrote, "A panel of 24 shots of a hand picking up a ball conveys a palpable sense of wonder, at both the intricacy of the action itself and the marvel of the medium that is capable of recording it in such a strange and powerful way." This "intricacy of the action" seems trivial now, with instant replays and slow motion, and the classic propaganda films on athletes as superior machines, Leni Reifensthal's *Triumph of the Will* and *Olympiad*. Muybridge's images were less importunate, and it seemed, unedited and comprehensive. Like C-Span, you could only stand so much of it at a time but wished there was more like it available.

Muybridge's most famous work was funded by railroad tycoon and senator Leland Stanford, who also financed the creation of Stanford University in Palo Alto. In a legendary bet, Stanford wanted to know whether all four legs of trotting horses left the ground at once. The bet was wagered in 1872, and there seemed no other way to prove it than by innovative photography. You could look at footprints in the mud, but a photograph would be irrefutable. Until the nineteenth century, many artists depicted horses that appeared to fly, with forelegs and hindlegs extended at the same time, like a rocking horse. *The Epsom Derby* by Gericault is an example, displayed in the Louvre. Muybridge was known for his photographs of Yosemite and developed a reputation for shooting more than photographs; his wife's lover met a bullet. His shots of horses showed that during a trot or gallop, all four hooves can be off the ground. Stanford won the bet, and several artists took note of the photographs and went back over their paintings to correct them.

The position of the horse's legs came as a surprise. A galloping horse sets down forelimbs, then hind limbs. They travel in pairs. A trotter moves diagonally opposite pairs at the same time. Muybridge's work revealed the changing positions during different gaits, and what's known as the duty factor, an obligation to gravity that, had he chosen slower horses to photograph, Stanford might have lost the bet.

The faster horses travel, the briefer is the touch of their hooves to the ground and the greater is the peak force on contact, necessary to carry forward the balance of the stride. The same principle is true for a human, and the peak force on a foot rises to 3.5 times body weight in sprinting. In jogging it's about 2.7. Neither humans nor horses are well designed for fast gaits. Horses have the better metabolic system but have weak leg bones for their weight. We expend more energy, which is partly why people jog, to burn fat, but our leg bones are stronger relative to body weight.

Muybridge's images have been described as "rigidly systematic," but equally comprehensive images of motion were captured by Etienne-Jules Marey (1830–1904), the French inventor and physiologist. One of Marey's earliest inventions was a "graphic inscriptor," used to measure blood pressure. He wanted to know how the body worked, and the camera was among many tools. (He invented the first film projector.) Marey is not as well known in the U.S. as Muybridge, but in Europe his innovations are heralded, especially his early work on birds in flight. He photographed birds with a camera that resembled a rifle, a loop of film rotating on the mount at high speed. The images that the rifle produced were three dimensional, and Marey used them to make models first of clay and then bronze, for the Science Museum in London, which also has a replica of his photographic rifle.

Marey began to work on his own stop-action camera following Muybridge's visit to Paris in 1881. The trick was to have the shutter open and shut at equal intervals without advancing the frame. Marey called his method "chronophotography" to indicate a time lapse. The result was the first multiple exposure, and it captured a pole vaulter before his launch, during his vault, and landing, all in one frame. Occasionally there were superimpositions in the ten or so images, and this influenced modernist artists, such as Marcel Duchamp in his *Nude Descending a Staircase*. Biographer Marta Braun reckons the impact of these photographs on art was "probably greater than any scientific work . . . since the discovery of perspective in the Renaissance." But his focus was the economy of motion, and Marey's portraits led to research in how to trim wasted efforts by humans at work and to refine physical training.

Marey was considered an amateur photographer, but he honed his methods for twenty years. His influence on the art world was a by-product of empirical investigations, like the work done at MIT by Harold Edgerton mentioned in chapter four. Using an exposure time of $1/50,000$ second, Edgerton captured the image of a milk drop on impact, with the corona of rising drops, and the latter phase of a splash, crater subsiding, center column rising. Marey's work on locomotion was published in *Animal Mechanism* in 1890, providing the earliest proof that pigeons can clap their wings above their back. This clap-fling mechanism in insects was captured on high speed film only in the sixties, by Torkel Weis-Fogh at Cambridge.

British photographer Stephen Dalton captures images of insects in flight, and when he began in the sixties, his goals were ambitious, to capture them in color, with high definition, in the insects' natural habitat. The necessary gear didn't exist, and he spent years perfecting the equipment. In the end he came up with triggers; the movement of a creature triggered a beam that triggered a lens shutter at $\frac{1}{400}$ second and a flash that fired at $\frac{1}{25,000}$ second. The normal shutter had a delay of $\frac{1}{20}$ second, "long enough for an insect to move ten inches away" and out of frame. The results were rejected by *National Geographic*, Dalton thinks, because they were thought to be faked. They have the look of paintings.

His multiple exposure of a lacewing featured on the cover of David Attenborough's *Life on Earth*, a sample of his own *Borne on the Wind*, published in 1975. Now his photographs appear in magazines such as *Natural History*. The disbelief continues with the cover of Neill Alexander's *Exploring Biomechanics*, which shows a basilisk, or Jesus lizard, sprinting across the water on its toes. I met Stephen Dalton in 1981, when working for Oxford Scientific Films, a group of British biologists who invented their own special lenses and probes. The filmmakers used high magnification to capture a compound eye and explode Radiolaria far beyond a tenth of a millimeter in diameter. Developing a method of time warp photography, they captured such things as a bumblebee in flight. They used a film speed of 5,000 frames per second to detail a wing beat of 200 times a second. Even with that frequency of upstroke and downstroke, the wings were seen to meet to form a single airfoil over their back. Their flight muscles have to reach a temperature of 30°C to achieve this frequency. So the bumblebee warms up before take-off and maintains heat by its own work.

Bees' flight muscles are located in the thorax, but key to their rapid wingbeat is the cuticle found in their wing hinges. Cuticle contains a highly elastic protein. The long-chain molecules assume different configurations. They move when affected, and their thermal motion has been described as Brownian. This rearrangement means the material can store energy and release it like a spring. The spring effect of cuticle works for a flea jumping, and we have a similar elastin protein in our ligaments and in the arterial walls of the heart. This long-chain molecular structure was copied for synthetic rubbers. If you stretch a rubber band across your upper lip, you can feel the heat.

"Exercise lies at the heart of the struggle for existence," according to the preface of a SEB journal, *The Comparative Physiology of Exercise*. As

Maggie Richards on *Body Electric* would say, Who would have thunk? The abilities of animals to exercise "are constantly being refined by the relentless process of natural selection." There are qualms with "constantly" and "relentless," but in a stretch you could say trees grow stronger with "exercise" as repetitive response to wind. Every limb is like an opposing muscle, stretching in tension on one side, contracting in compression on the other. Fish, birds, sea turtles and whales migrate for much longer distances than terrestrial creatures, which expend more energy doing so. There are always exceptions; a hummingbird expends enormous energy. Power requires other adjustments. A rabbit has a vascular hydraulic system in its head to prevent injury after a big jump.

How do you put a price on the economy of motion? David Attenborough noted that many birds became flightless when isolated on islands with no predators, because of the tremendous energy that flight requires: "Birds appear to abandon flight whenever possible." Others, like Charlie Ellington, say flight is economic because of the velocities achieved. Certainly it presents an option, which the dodo failed to exercise. Most flying birds have relatively larger hearts than mammals.

It was in the mid-eighteenth century that it was first discovered that animals that endured heavy exercise had increased muscle mass and larger hearts. This seems late in the game, since Leonardo da Vinci studied the heart, and hunting and fishing were so common that anyone who prepared an animal for food would notice this—especially considering preferences for lean meat, or white or dark, the latter denoting aerobically exercised muscles. But comparative measurements were only begun in the seventeenth century, when Giovanni Borelli published the first textbook on biomechanics, *On the Movement of Animals*. Much of the book is devoted to how muscles work, and his experiments included testing arm muscles. Borelli devised a formula for measuring their strength, from biceps to hand. The total adds up, but as Neill Alexander pointed out, Borelli did not know of the intricacy of muscle fibers and misjudged the segments at work. He could not see them.

Borelli "got into deeper trouble," when he tried to measure the forces in a standing jump, Alexander wrote. "He started with a static problem calculating muscle forces for a man standing with legs bent, as if about to jump. He then tried to move to the dynamic cause by an argument that confused force with momentum and reached the spectacular conclusion that the total force was 2900 times body weight." Modern calculations give peak forces of about 8 times body weight in the knee extensor muscle, using a formula familiar to Alexander.

R. McNeill Alexander is a professor at Leeds University in Yorkshire, and a pivotal participant in both SFB and SEB. He describes the sarcomeres that Borelli couldn't see, muscle filaments that create cross bridges, thick ones swinging across and attaching to transfer the force to thin filaments. The number of filaments in a sarcomere influences its strength, but the big factor is length: the longer the better. Humans gain no particular advantage from being large; the fastest muscle that has been found is attached to a rat's eye, and the most powerful muscles are those that move the wings of insects and hummingbirds. Alexander not only has a grasp of the intricate movement of muscle fibers, but he looks at animal skeletons as structures, calculating safety factors for bones and measuring the economy of spring action, data which led to refinements in running shoes and indoor jogging tracks.

In his studies at Cambridge, Alexander was influenced by the pioneering work of James Gray. Like D'Arcy Thompson, Gray lectured in the annual Christmas lectures for children at Britain's Royal Institute, yet the simple models he built are useful for adults, since the facts of locomotion can seem counterintuitive. ("Muscle fibers do work by shortening while exerting force," wrote Alexander. "A muscle that is extended while exerting a force has work done on it, and degrades this work to heat. Muscles do negative work when an animal slows down.")

Alexander still uses the models he inherited from Gray. One is a fish of foam rubber; when a spring is released, the tail moves sideways and the fish moves forward. A segment of a snake is represented by a curved rail; in the inner curve, muscles contract, then expand. All of this appealed to Alexander, whose father was a civil engineer and whose mother wrote children's books. After studying zoology at Cambridge, he taught at University College in Wales, where one of his students was

Meave Epps Leakey, now head of paleontology at the Nairobi Museum. "His explanations were always so clear," Leakey recalled. "I learnt more from him than anyone else. As a lecturer, he was unquestionably the best in the department, illustrating his points with entertaining models which clearly showed the mechanical principles, or using his own acrobatics."

Alexander's lanky six-foot-four frame made use of a bicycle on campus; he was a moving sage with a white beard, which he acquired long before his current age of 61. Pedaling a bicycle requires half the energy of running. Alexander measured a 1:2 ratio of oxygen consumption for the two activities. Every cubic centimeter of oxygen that we use releases about 20 joules of energy from our body. People standing still use energy at a rate of about 120 joules per second, but a jogger's rate increases by a factor of ten.

The mechanical principles of a human running "are the same as those of a kangaroo hopping," Alexander says; "Runners bounce." Ninety-three percent of the energy is returned by the coil action of the Achilles tendon. Tendons are not as elastic as rubber; like a rope, they break when stretched over 8 percent of their normal length. But even that small give is enough to save leg muscles a considerable amount of work. When running, muscles act like brakes against the ground during the peak force, degrading energy to heat, the same way car brakes usually dissipate energy. But the spring action of the Achilles tendon trims the loss to 7 percent.

Alexander began to look for other springs, and studied films of barefoot runners, focusing on the arch, which is unique to humans. Several other primates move about bipedally from time to time, including baboons and chimps. Even when they're on all fours, they carry most of their weight on their hind legs. Also, they're flat-footed, placing the heel as well as the toe on the ground when they walk. Bears are also flat-footed; cheetahs and horses stand on their toes. It's easier to carry more weight, and maintain balance, with the full foundation of the sole as support. The habit of setting the heel down before the toes seems unique in humans; consequently, the heel became more prominent because of the pressures on the bone from an increased load. This enlargement of the heel combines with the pad behind our toes to create the arch. A barefoot runner's arch goes totally flat when it hits the ground, but it also works as a spring. Alexander could measure the effect by using a foot plate and loading with pressure severed feet, courtesy of car accidents and other misfortunes. When we discussed this experiment in a British pub, I switched from the local ale to a scotch and caught myself rotating my ankle.

He found that, together with the Achilles tendon, the arch halves the work of leg muscles. "Our experiments on the arch of the foot were widely reported in the popular press and in running magazines," he explained in a lecture. "About a year later many manufacturers started emphasizing in their advertisements the elastic properties of the sole of their running shoes. They promoted 'energy return' soles that are not unusually thick and do not compress to any unusual extent." Alexander tested the shoes, loading them with pressure, and found they returned only 66 percent of the energy, less than the natural spring of the arch provides. "They may nevertheless be helpful. At each footfall, about 8 joules work are done compressing such a sole and it returns about 5 joules in its recoil. If one brand returns even 1 joule more than another, that may give a significant advantage in competition."

Tom McMahon used the spring principle to refine a running track at Harvard. The track is described as "tuned," and returns some energy to a runner. The original plan was for a concrete base with ³/₈ inch of polyurethane on top. In collaboration with Peter Greene, McMahon set out to design a more compliant surface, creating a mathematical model on the computer and then filming runners on various surfaces using high-speed photography. Ground reaction forces were measured by foot plates. The atheletes said it felt like gliding on air, as if they were running just above the surface. The new surface resulted in trimming five seconds off the mile and it reduced injuries. The Harvard coach found he could train the team harder than before but still reduce injuries by 50 percent. The track was economic in other ways, costing less than the original construction planned.

The concept of gaining energy from a spring is hardly new to sports; a pole vaulter stores kinetic energy in the pole, which releases that energy, and the elasticity of fiberglass increases the height of the vault. A fishing rod has a double helical bond, allowing response and strength. But now the momentum has turned to other fields, to create shock absorbers that respond to local road conditions without a computer, and more fatigue-resistant parts for aircraft based on the pattern of fibers found in a beetle's shell.

A 1990 conference in Tucson, Arizona, focused on Smart Materials, including material that enables concrete bridges to self-repair. "Nature has had millions of years of evolution to perfect things," said Paul Calvert of the University of Arizona. "We would like to look at biological structures and try to transplant them into synthetic systems." This involves several tricks. For example, merely padding a running shoe with bubbles in the sole can lead to imbalance and injuries. The approach has to be tested and refined; the heel of the foot takes the brunt of peak force impact. Also, the mechanics of dynamic structures are revealed in an evolutionary view, the same way that aviation went through phases that resembled the evolution of natural flight. So it's useful to know about how the arch of the foot was shaped. Because organic structures are multifunctional, investigators see different things at different times and priorities change.

There are many springs at work in locomotion. Antelopes have a spring in their ankle bone, which allows a rotating movement at both top and bottom. This double-pulley ankle enhances the animals' ability to leap and bound, a capacity refined by springbok and deer. A flexible back, working like a spring, seems to contribute to the power of cheetahs, said to achieve speeds of 70 mph. But the cheetah's burst of power is not maintained for long. Sustainable power is produced aerobically, and is twenty times more efficient.

A hummingbird has the highest measured rate of aerobic metabolism. When hovering, some species consume 80 cubic centimeters of oxygen per gram of body weight per hour. Just before their migration, which may cover 3,000 kilometers, they change their diets to carbohydrates to build up fat, gaining about 10 percent of their body mass for each day of their premigration stuffing. En route, they make nectar stops only. Burning sugar from nectar saves their storage of fat. Their liver and muscles contain only enough nectar fuel (in the form of glycogen) to power five minutes of flight, and hovering is the most expensive form of flight. So they keep their nectar stops under five minutes; otherwise they will dip into their fat reserves, which they use in their natural flapping flight. The nectar replenishes the glycogen reserves, moving directly into the system courtesy of an intense metabolism. Hummingbirds have high rates of absorption and rapid synthesis, but there is very little safety margin in their reserves.

Birds of all sizes have a better respiratory system than people, because oxygen runs throughout their entire system of auxiliary air sacs and can enter and exit, as mentioned, even by their toes. The result is a constant flow of new oxygen, while in humans and other mammals stale air is recirculated with the incoming mix of fresh air.

The form of a fish face is conducive to high oxygen intake, with a flow directed right into the gills. The blood of the fish and the water flow in opposite directions, so, like birds, fish run into air rather than running out of it. Blood in fish gills circulates in an exposed part of the body, so the blood cells expel the stale air, then retrieve the highest possible intake of a fresh supply, without it being filtered through the system. With this kind of flow, fish extract around 80 percent of the oxygen in the water. Oxygen extraction in mammals is only about 20–25 percent.

The efficiency of a horse's respiratory system is relatively high; oxygen flows faster through pathways to the blood cells and directly to the mitochondria, and is absorbed at a rapid rate. Horses breathe more economically than humans, cows, or dogs. The value of oxygen to stamina is such that breathing is coupled in the same phase as the movement of the limbs. This phase locking occurs when kangaroos hop, when dogs bark, and when bats emit their echo locators. When running, a dog barks when its forefeet hit the ground. Bats send out their echo signals in sync with each wingbeat, emitting a signal when exhaling on the downstroke. Birds exhale on the downstroke, compressing the rib cage. A nautilus also couples breathing and locomotion.

Jet propulsion is one of the most ancient forms of locomotion. Nautilus fossils have been recovered from deep sea deposit horizons laid

down during the late Cambrian. A nautilus has two methods for producing jet power. The first, its original MO, is simply to withdraw into its
shell rapidly, which forces all the water out of its mantle cavity in a spurt.
Because a nautilus secretes gas, which rises into the top of the shell, the
animal is buoyant. Smaller chambers are filled with nitrogen. With this
buoyancy, and a rapid flush of water, a nautilus rises off the bottom of
the sea floor.

What may have begun as a way to simply hide from predators became
a means of escape, which developed into a way to travel. As Dolf
Seilacher has pointed out, every evolutionary breakthrough opens doors
that may not have been intended by the original function. Why crawl
when you can use a jet?

The secondary method for producing jet power came from little
wings and muscles that developed to increase the velocity of water flow.
A foot once used for crawling developed a series of flaps that pushed the
water out. Interior "wings" direct water intake for the gills. These wings
funnel with a wave of muscle contractions—springs again. They continue to work, but less vigorously, when the nautilus is just cruising in
the water or at rest. This slow and easy flow of water allows the gills
maximum time to extract the most oxygen with minimum effort. So
the respiratory system can become more efficient as the oxygen content decreases. A nautilus manages to consume enough oxygen to flourish in waters that most fish would find suffocating. The wings contribute to thrust when it's time to jet, joining with the flaps that direct
the flow.

How this breathing and locomotion coupled is explained by Martin
Wells, who studied zoology at Cambridge. The delicate wings of the
funnel alone exert only enough pressure to move the nautilus along at
about 3 or 4 centimeters per second. To gain speed, the nautilus used its
old defense ploy, rapid retreat, exercising retractor muscles that are
described as massive. Their power increased the speed of the nautilus
ten times.

"Thirty centimeters a second was doubtless spectacular early in the
Paleozoic era," Wells wrote, but "it didn't look so good by the end of the
era." This is when fish appeared on the scene. Fish were neutrally buoy-

ant because of their pneu bladders, they had jaws, and they had a more efficient form of locomotion. Thrust is measured as the mathematical product of mass and velocity. But the energy cost required to push water rises as the velocity squared. So it's energetically cheaper to push a large mass of water at low velocity than it is to accelerate a small amount to a high velocity. A fish can sweep back a volume of water greater than the body mass of the fish itself.

Other creatures use jet propulsion, including squids which experimented with their shell, and in the end, simply grew out of it. The only thing left is a long thin rod of chitin that supports their stomach. They can expand more, but they lose their buoyancy and must swim constantly. At slow speeds, all their energy goes into staying off the bottom rather than moving forward, and at high speeds they use four times the energy that fish do.

"Having burned their boats," Wells wrote, "squids exploited a mechanism that is widespread in planktonic animals." They altered the ionic content of their tissues by hanging onto ammonia and displacing sodium ions. Ammonia is lighter than sodium, which made their tissues pneus against salt water. They not only gained buoyancy but could achieve it at greater depths.

In its own experiments with the shell, the nautilus trimmed weight. This explains part of the beauty of the logarithmic spiral. By growing from one end only, and by making additions of real estate as they grew, each chamber was built on the foundation of the previous home. The nautilus is the only one in its family to maintain a shell. With its own buoyancy in the top of its shell, at slow speeds it remains as energy efficient as a fish.

Observed Wells: "The jet set, after tinkering with the motor for millions of years, now seems to be doing its best to develop less expensive means of getting about. There must be a lesson in there somewhere."

◆ ◆ ◆

In the early days of the Society for Experimental Biology, the focus was on medical implants and spare parts. Efforts were international, and the results familar. Balloon catheters are used to clear blocked arteries, and artifical valves have been implanted, with limited success. Some were not as fatigue resistant as the screw Claus Mattheck designed, and their little struts cracked. Hips were often replaced by stainless steel, taking the load off the attached real bone, which wasted away. (The reaction to decreased loads is immediate. In 1994, one of the space shuttle astronauts reported that his spine had elongated 2 inches after only two weeks on board.) Now the materials for hip replacements are being made more organic and lighter in weight so as to integrate with the host tissue. This kind of organic implant is a focus of research at the University of London. The same approach is being taken by physicians in New York with torn knee cartilage; the cartilage is frequently removed when damaged, but the procedure can lead to arthritis. Now the emphasis is on healing by inserting a graft. Japanese researchers are testing artificial cells that imitate healthy pancreatic cells for people who suffer from diabetes.

In Japan, research into so-called Smart Materials began in 1987, funded by the Science and Technology Agency, and the focus included ceramics, mechanical engineering and pharmaceuticals. One of their projects investigates materials that can sense their environment, thin films of chemicals with optical properties. They simulate nerve impulses and are visible in laser light. The piezoelectric effect, common in bones and muscles, has been copied into crystals at the Grumann Research Center in the U.S. Inserted into airplanes and automobiles, they could work to dampen vibrations. Piezoelectric elements have also been proposed to reduce drag by allowing wings to warp in flight. Fiber optics serve as nerves in the material of aircraft and respond to changes in temperature and strain, which can be read by optical signals. The frequency of light traveling down the fibers serves as an advance warning of subtle changes before damage occurs. Alloys are being designed to have a shape memory, that is, to return to their original shape after being exposed to heat.

Flexing their backs like cheetahs, dolphins and whales tend to swim by moving up and down, rather than using the side-to-side motion of fish. Using these springs as a model for submarine material, researchers at the

University of New Mexico have designed polymer gels that stretch and contract like muscles when their backbone of polymer is charged or neutralized. At Marsten-Bentley in Houston, polymers are being used as a seal called Hylomar. Originally developed by British scientists at the Rolls Royce aerospace divsion, Hylomar was used to coat jet engines in the sixties. It quickly became the duct tape of perpetual putty, used to make emergency repairs on racing cars to hold things in place temporarily. Now new structures of this polymer are being developed, with molecules that branch from the center like trees. Each ray acts like "a sticky tentacle," and the sealants are so elastic that they squeeze foreign materials from concrete joints. At Marsten-Bentley, the new structure of Hylomar has helped simplify product designs. A tight seal, for example, got rid of the aluminum and rubber trim around car windshields and created a better bond with the rest of the car in case of an accident.

Ina Goldberg, a geneticist at Allied-Signal, studies the adhesive that mollusks use to attach themselves to rocks, to try to duplicate it in a waterproof glue. At the University of Reading, Julien Vincent has created a fiberglass resin based on his studies of the composite structure of wood cells. Researchers at the University of Washington have refined a material that resists impact, based on the calcium carbonate shell of abalone. The shell of a horned beetle contributed a blueprint for lightweight but strong aeronautical material. By studying ivory with a scanning electron microscope, researchers at Rensselaer Polytechnic Institute in Troy, New York, developed a synthetic that duplicates the feel of real ivory, preferred by pianists. Polymers have been used for piano keys before, but it was considered slippery and cool to touch. The scanning electron microscope exposed the natural pores and random peaks and valleys that occur in real ivory, and this was duplicated. The first attempt to duplicate ivory was in 1869, when a New York inventor named John Wesley Hyatt came up with a synthetic substitute for ivory billiard balls. The substitute didn't have the right density or elasticity for billiards, but other uses were found for celluloid.

Caroline Dry at the University of Illinois developed fibers that release a compound glue to heal cracks in concrete as they occur. The hollow fibers are distributed throughout wet cement, and, when flexed, work like a time-release capsule, supplying an adhesive to fill in the crack. In addition to safety factors, such self-repair would reduce maintenance and prolong the life of concrete bridges, roads, and any cement structure. The concept is similar to reaction wood and the way bones repair a crack with mineralization. Professor Dry's work was funded by the National Science Foundation.

Dr. Dry delivered an early report on her research at a 1991 conference on Active Materials and Adaptive Structures in Alexandria, Virginia. Japan also hosts an annual international Workshop on Intelligent Materials. The University of Arizona hosted a conference on Materials Synthesis Based on Biological Process. The Materials Research Society meets every year in Boston. The first European conference on smart materials was held in Glasgow in 1991. A Smart Structures Research Institute had already been established at the University of Strathclyde in Glasgow. Tokyo University has a Department of Material Systems Engineering. The University of Toronto has a Fiber Optic Smart Structures Laboratory. Michigan State University has the Intelligent Materials and Structures Laboratory. The *Journal of Intelligent Material Systems* was launched in 1989. In 1992, a quarterly journal called *Biomimetics* began, a joint project of Julian Vincent at the Biomechanics Group, University of Reading, and A. V. Srinivasan at the Applied Mechanics Research division of United Technologies in East Hartford, Connecticut. DuPont and McDonnell Douglas have divisions devoted to Smart Materials.

Craig Rogers, a mechanical engineer and director of the Center for Intelligent Materials Systems and Structures at Virginia Tech, described Caroline Dry's concrete repair fibers as "bio-mimicking." Ronald F. Zollo, chair of the American Concrete Institute subcommittee on reinforced concrete, said the research "sounds interesting as a concept," but warned that the economic practicality would have to be studied.

John H. Perry, Jr., is the head of Energy Partners, based in West Palm Beach, Florida. Part of Perry's fortune derived from designing submarines for offshore drilling, but now he thinks the hydrogen fuel cell is "the silicon chip" of the future. He built the Hydrolab, used for training astronauts under water for weightless conditions, and is developing a car using hydrogen cells. Hydrogen is easy to produce by combining solar power and water. Such a power plant exists near Munich; Mercedes-Benz and BMW have designed prototypes that run on hydrogen fuel. New hydrogen cells have been developed to produce electricity. When the same amount of fuel is converted to electrical power, the cars can travel twice the distance. If Perry's hydrogen fuel cells were combined with the aerodynamic design of Paul MacCready's Impact, the range would increase to about 400 miles without recharging.

In 1993, General Motors solicited volunteers to test the Impact. Applications went out in electric bills. To be selected, the drivers needed a garage where a charging unit could be installed for a month. They would pay only for the electricity, which costs less per mile than gasoline. GM expected 4,000 responses in Los Angeles but received over 9,000 applications. In New York, the company received 14,000, and it stopped the offer a month earlier than planned. During an auto show, a GM spokesperson said the company "is prepared" for the Impact to fail. This struck Tom Jorling as an "unbelievably curious" position. Jorling is head of the New York State Department of Environmental Conservation. New York, like California, must comply with new emissions standards; by 1998, at least 2 percent of all cars sold must be electric. Jorling pointed out that consumers have demonstrated tremendous interest, yet Detroit tends to rejects the new technology. He suggested the real reason is that automakers don't want their investments in the combustion engine technology to become obsolete. "It is not going to help if we build 'X' number of units to satisfy the mandate, and then we don't sell them," explained a GM spokesperson who emphasized, "Clearly, the consumer is going to decide the future of the Impact."

9. The Greater Flux

IN *The Man in the White Suit*, Alec Guinness plays a young scientist who invents thread of such a molecular consistency that it won't break. His white suit will never develop holes in the elbows, nor will the fabric stain. Union leaders within the clothing industry react to the discovery with horror; tailors would be forced to retire their needles, dry cleaners would go on welfare, and fashion is, after all, designed to become obsolete. With a fatigue-resistant wardrobe in the closet, the bulk of the British could scarcely be persuaded to return to Bond Street for a trendier frock.

But the British film director Alexander Mackendrick managed to blur the lines between the good guys and the bad. The inventor is attacked not by the predictable villains of a paranoid industry but by the very public that was to benefit from his discovery. Cornered, Guinness becomes the mother of his invention, rappelling down a wall, using his thread the way a spider dangles its body weight by a dragline. But it's hard to hide in a white suit, and the garment disintegrates when the mob grabs him, clutching pieces of white cloth that fly into the air and float to the ground as if they were "plucking feathers off a chicken," as a critic for the *New Yorker* wrote. Other films by Mackendrick were mentioned in a tribute to the director that followed his death in 1993, including *Sweet Smell of Success*. It was said that the world Mackendrick "spins has two sides, the bright and the bitter—and it never comes to rest."

Such a world of dynamic instability figures in the work of Peter Allen, who was featured as the guest speaker at the 1991 SFB Natural

Structures conference in Stuttgart. Allen talked about the theory of self-organization, which Dolf Seilacher sometimes calls chaotic self-organizing. It embraces the notion that fluctuation can lend order, or build patterns. It also expresses the synergetics that Bucky Fuller promoted: the behavior of the whole is unpredicted by its parts. Finally, there is the twist that changes involve a process that cannot be reversed.

D'Arcy Thompson was sitting right on top of the theory when he wrote about generating patterns, an unfurling with underlying principles. Or a building of snowflakes and crystals, with underlying principles. Ink drops in water, irretrievable. Hexagons with principles. The principles were inevitably linked to physical laws.

Allen's approach is partly explained by his previous work as a physicist. "It's all very Buddhist, what I'm saying," he reflected during an interview at his office in England. "It's accepting the basic incompleteness of any knowledge. The fact is that we can never know. All descriptions are necessarily incomplete." At Stuttgart, he had said, "You don't know what you don't know."

Were I not in possession of a nonlinear brain, I might have folded up my notebook and called it a day. The circle of Willis, located at the base of the brain, is described in *Gray's Anatomy* as a "remarkable anatosmosis." It equalizes circulation in the cerebral arteries, and makes provisions for carrying on if any of the branches are obliterated. That it was named by a British physician, Thomas Willis, is coincidental to this work. That it is circular and capable of self-repair is not.

As a letter of the Greek alphabet, delta is synergetic in that it can be organized to form words, but it can also be discrete, a symbol of change within formulas. A triangle may serve as the lucid blueprint of a structure of strength, but the formalized delta was based on the flux of meandering tributaries at the mouth of a river, and the greater flux of alluvial sediments that a flood imbues, outside the boundaries of predictable flow. There you have it, the theory of self-organization. It can fit almost anywhere, even swallowed internally.

Allen's expression of the inchoate is meant to acknowledge the reality of change. It is the Houdini at work that Orville Wright acknowledged when writing of the tricks to flight—now you see it, now you don't. British biologist Rupert Sheldrake has even proposed that physical laws, such as those of gravity, evolved, and may change again in the future. Some experiments in nature suggest that for every law there is an outlaw, but often the culprit is the human observer, overarching with formulas, creating concepts of what he or she wants to see, or just being downright wrong. With every revision, people's concepts have to change, and there are times when you just want to say Forget it, or that single word that expresses a policy statement for many people now: Whatever. Bucky Fuller saw this dilemma as a legacy of the Pythagoreans that stretched "to Einstein himself, that one could expect tidy relationships of numbers in the very depths of the universe, and the tradition, seldom dwelt on but always obscurely suspected, that scientific thinking at a certain point will always stop making sense." For the Pythagoreans, the latter included irrational numbers. For physicists like Freeman Dyson, it is discovering that the facts of matter seem to be increasingly counterintuitive.

Newtonian physics explained the world as working like a machine. Parts interacted like cogs and wheels and gearboxes, and the Creator was described as a blind watchmaker: the mechanism was set in motion and it kept on ticking. In the absence of friction, movement—such as that of the planets—was infinite. With friction, motion was dissipated into random thermal motion, heat—thus the idea of entropy and the prophecy that the universe will wind down. Thermodynamics was so widely successful that its inherent concept of equilibrium was applied to economics and evolution. But chaos and the discovery of random changes, moments of punctuated equilibrium, altered the notion of steady progress. Newtonian physics explained a lot of things, but it didn't explain evolution.

D'Arcy Thompson underscored the influence of physical forces, but this did not explain random choices, the alterations of parts within the machine. The squid chose one pathway to great depths, the nautilus another. Apparently living things are just as restless as the earth's surface, and we are fooled by the illusion that we are on solid ground. This is the enigma that Allen embraces. He is not alone; at Stuttgart he was given the reception of a visiting apostle.

To many thinkers, changing patterns in nature and evolution provide a more realistic model for economics and social systems because they, too, are dynamic. The theory of self-organization is appearing in disciplines where you would never expect it, in perfect imitation of itself. At Stuttgart, Allen presented a holographic model of an evolutionary tree, produced by a computer, and he showed it on a video monitor. The image did not resemble a tree until the final stages but began as a tiny blob that was more cosmic-looking than earthbound. It grew with changing hues to suggest instability, different colors fading in and out, and branches and species concentrated in the center, but some "coming out of nowhere." Allen pointed to clouds that condensed and formed a lineage. They resembled the red sponge in the first stages of rebuilding itself. These clusters appeared in the blank but broad margins of the form, which Allen called a "possibility space."

"My simulations are the first ones ever done that actually take what Darwin said and do it on a computer. You end up with an ecology with diverse behavior. It's stable only when the explorations are repressed all the time by selection." An exploring species would be one like the squid, which having relinquished its shell, devised a better way to reach greater depths with its ammonia trick. Three million years ago, humans were doing some exploring. "The whole species has stopped exploring a wide range of morphology. But as soon as you do something, like alter the environment, it will automatically respond in a very complicated way which is not mechanical. Species will adapt, and change. That's why evolutionary biologists can't even define species. They change, and what is it?"

Our ancestors 3 million years ago were not considered human but merely hominid, upright primates. Many experts think the change to a bipedal stance was influenced by a changing environment, a transition from dense forests to savannahs, with only pockets of trees. Defining species within our own lineage has been controversial. A gracile australopithecine or *Homo habilis*? A plant or animal, or a zoophyte? An Ediacara weirdo or just a sponge in the making? Even a banana tree is not considered a true tree, but a giant herb.

"If you go to a biologist, he or she won't even be able to define a species. If you look at my simulation you'll see why: initially species are not species. The notion that something isn't, then is, cannot be dealt with in the traditional way of thinking." Allen also means a time in the cosmos when there was no time, life from nada, like those clouds in the hologram that came "out of nowhere." He complains: "Science will say there must have been something different internal already. It's like the Russian dolls."

The Russian dolls embrace the notion of similar things within smaller things—specks of dust suspended like planets, on this planet, among other planets, within this galaxy, within this universe. Some scientists see a teeming universe in a drop of water, drops of water within the ocean, and so on. As Jonathan Swift wrote:

> *So, Naturalists Observe, a Flea,*
> *Hath smaller Fleas that on him prey,*
> *And these have smaller Fleas to bite 'em,*
> *And so proceed ad infinitum.*

Before the Big Bang, the cosmos was presumably empty but for a convention of quarks. The earth was devoid of life for billions of years. And land, such as it was, began barren. Says Allen: "If you have an empty system, a particular species will grow up very successfully, because it puts its behavior in itself. That behavior is amplified, until it finds a limit, then it explores new space, and what was punished before as an aberration develops in an empty space. Whereas being different before was punished by lack of success, once you hit that limit, being different is now rewarded." Whales, for example, among the mammals that returned to sea, were perhaps an aberration on land, but a success in the ocean. Allen's next analogy belongs to sand dollars, changing to tap the food resources on the beach. "In a competitive situation," he says, "they drift out until their behavior is sufficiently different from the ancestral one, to a point where they tap into a new resource. At that point, they can multiply. So the system fills with separated species."

Allen applies the analogies of evolution to everything. "When you think of the human brain, diversity and exploration, individual liberties, all those things fit this picture. Evolution is shown to select for populations with an ability to learn, rather than for populations with optimal behavior. Change is driven by the nonaverage." Deviants are normally suppressed by the system, he says, but at certain times, the system cannot suppress them, "and the deviants restructure the system." The French Revolution comes to mind, but sand dollars have also been accused of breaking the rules. Allen thinks that the "long-term survival or extinction of humans, like any species, is related to its ability to cope with uncertainty and change, to generate responses." As Richard Leakey once said to me, "If you don't make a decision, other forces, and other people, will make it for you."

Allen's argument for a dynamic approach is the same one that Dolf Seilacher made with Morphodynamics to try to free the study of forms from being static. "The point is," argues Allen, "that in terms of description, in terms of these variables and how they change in time, normal science can't deal with one thing that changes into another." What he does is not normal science, but apparently it is part of one of those clouds gathering and condensing into something.

The targets may be fluid and nonlinear, but Allen's projects are beamed toward profits in a changing marketplace, including the design, in conjunction with Lotus and Volvo, of a prototype vehicle with a lean burn system intended to average 200 miles a gallon. He says they didn't want to go electric, "a total revolution in the system," but wanted simply to modify the current mode.

Other projects with the European Economic Community (EEC) concern impact studies, research into sustainable practices in agriculture and the fishing industry. His model of evolution is applied on a small scale to specific projects in city planning and recycling. The theme is resource management, and it folds in the needs of people to a remarkable degree. The activities of human populations are not separated from the landscape.

◆ ◆ ◆

PETER ALLEN is a research professor at the Cranfield Institute of Technology, located in a cow pasture near Bedford, England. In 1987, he was selected to head the Eco-Technology Center at Cranfield, which is the largest postgraduate center in Europe, with 3,000 postgraduate students. The center was built on a grant from the Honda Foundation.

Sochira Honda died in 1991, only a few years after he gave £1 million to establish the center. The Honda Civic was one of the first successful economy cars, with exceptional gas mileage. Allen describes Sochira Honda as "a very original chap, who was keen on taking the environment into account. He felt that technology developed on cost basis alone was very destructive. He also devised a code of practice that I use myself"; he pulls down a copy of the *Honda Book of Management* from the bookshelf, and says that he considers it more sustainable than *In Search of Excellence* or anything else written on reinventing the corporation. Many recognize the form of bureaucracy itself as a constraint, yet others see evolution operating with a hierarchy, from cells, to organs, to organisms, to species. How then, do you develop little clouds of possibilities on the side? Diversify? For the moment, one solution is the interdisciplinary approach, creating little pockets of hybrids. Some may turn out to be mules, but that is the sort of risk that Allen sees as an essential part of exploration. Another way to break the constraints and costs of a bureaucracy is to pull in free-lancers, which many advertising agencies have begun to adopt, to trim their overhead, hiring people on a project-by-project basis.

Honda wanted the endowment to go towards research of ecology and technology, side by side. "This center is really unique," says Allen. "I'm always amazed, because it seems this is so obviously what we should do." The Sante Fe Institute undertakes similar computer models, "but they focus on physics, and complexity—not policy." Many companies, such as AeroVironment, combine environmental research and technology, but they may pull in outside expertise in science for specific projects, like the Q.N.

On the new white walls of the Center, some of the research is framed as art: hexagons, fractals, the crystalline, strange attractors, waterwheels, lines of numbers, the random. "These begin as mathematical structures," Allen explains. "We apply sets of rules to the patterns to see if they develop cycles eventually. If you have a fixed rule, and you have a series of zeros and ones and repeat that on the next line, the question is whether you can return to your original condition, or whether you can get on many separate cycles, or just one big one. You want to discover what the different possibilities are."

D'Arcy Thompson described similar experiments in pattern formation, called diffusion models. He referred to several works around 1900 in which thin layers of liquid were heated in a copper bowl. With liquid heated in an unstable condition, one wouldn't expect neat patterns of symmetry. But if the heat is kept uniform, patterns begin to organize. The first stage can be long tubes; then right angles are formed, at even distances, and result in hexagonal cells. In other liquids, at other temperatures, the initial form is circular, creating spiral curves. Thompson referred to the intermediary phase as one of partial equilibrium.

Allen's most influential colleague was Iily Prigogine, who received the Nobel prize in 1977 for his work in thermodynamics. Prigogine based his observations on experiments in physics and chemistry similar to the ones Thompson described, and called them dissipative structures.

A laser is a common example of such dissipation. Ink dropped into a glass of water dissipates, spreading evenly, but the reverse is never observed, like gas escaping from a pressurized tank into the air. This growing disorder was considered a given of entropy, the dissipation of kinetic energy, as happens with a car brake, or your leg muscle extended in tension. But Prigogine emphasized the unpredictable—that in a closed system, where energy and matter do not flow through boundaries, a system can exhibit spontaneous self-organization. The result was nonlinear equations for these systems of interaction. The equations can be applied to computer "art," but they can also be applied to any system that might be considered synergetic, such as economic, urban, and ecological systems.

On a small scale, these patterns of self-organization pop up in biology. The cells of sponges managed to reorganize themselves, as does the freshwater polyp known as a hydra. Cut in two, a hydra forms two new animals. The one with a head generates a foot; the one with a foot grows a new head. Seilacher thinks that self-organization is behind the common generating patterns of zebra stripes, fingerprints, and the generation of soft and hard components.

In his book *Exploring Biomechanics*, Neill Alexander describes small-scale locomotion among flagellates, organisms so tiny and so bottom-heavy that any turbulence in the water has profound effects on them. In a simple experiment, a glass beaker full of these tiny creatures was stirred and left to sit. At first the liquid was uniformly green, but within minutes, vertical streaks began to develop. The flagellates had begun to concentrate, swimming upwards in groups. Because they are denser than water, the upper layer of suspension became denser than the rest, "an unstable situation." So groups started sinking, and a phenomenon called hydrodynamic focusing drew them together in plumes. The focusing is partly a result of viscosity and partly the result of their shape. The water was set into a circulating pattern, with sinking plumes. The final result was like the convection currents in a heated dish. The process is called bioconvection; you can see it in the surface of water with a dense population of algae. If the water is disturbed, they will go into an unstable condition, then reorganize themselves.

"Self-organization was starting to appear in 1969," Allen recalled, when he first began to work at the department of physics at the University of Brussels, where Prigogine was based. "They were just beginning to work with what was called a Brusselator, named in honor of the location. It simulates a chemical reaction which is sufficiently nonlinear that it exhibits all these properties of self-organization. They were discovering all these things during my first years there. Everybody outside, especially in the U.K., thought it was all wild theory, that Prigogine was a bit of a fraud. Of course, that changed afterwards."

Allen stayed at Brussels until 1987, working on viscosity. "It was macro behavior from micro details, drawing from Prigogine's work. But the problem seemed a bit boring, and I wanted to get into these other areas, of ecology and social sciences, but take his theories with me." At the University of Brussels, Allen developed a group that included geographers and anthropologists who "spent some time educating me." With their input, in 1975, he began using a holistic approach to problems in urban planning, applying the theory of self-organization to cities and flows of transportation. In 1976, his group, working with the University of Texas, was granted a five-year contract to conduct research for the U. S. Department of Transportation, or DOT. "That was wonderful," he recalls. "My group was really initiated by DOT money." After Ronald Reagan left the office of the presidency, the funding dried up for his group. "Research grants went to firms that were in some way politically connected. It became a closed circle, even with NSF," the National Science Foundation. "It's not easy to get funding for genuinely new ideas."

At first, the Honda Foundation "didn't know what they wanted to do," he recalled. "Would it be a conference center, with visiting scholars and designers from Japan? We came through with the idea of a research center, which is what Mr. Honda wanted. We're externally funded now; outside projects bring in 60 percent of our finance. We've got twenty-eight Ph.D. students operating from our center. It's all research-oriented. We don't teach courses. We get the students involved in our projects, and they do their thesis on a related aspect. Initially we were interested in the technology side of change, and now we've made it more ecological/ environmental side."

Most of the research contracts are with EEC, UNEP, and private corporations. The British government provides a purse for each graduate student, and an anonymous donor ("a wealthy millionaire") contributed for the "fundamental work on evolution."

Once a project is commissioned, part of the Center's purpose is to bridge gaps between disciplines. In the case of a study of soil degradation and desertification, for example, Allen says, "the EEC may send a scientist to study the soil, but in reality, that science cannot translate into a policy unless you understand why the farmers are doing what they're doing. Without the broad study of the economics and the ecology, you can't really make policy decisions. Most policy isn't sensible, because they operate at a micro level, some edict which generally has the opposite effect of what was intended. We try to show what the behavior of the system may be in the longer term."

The Eco-Center provides the glue to put people together. To help find out the effects on local people, students from the Eco-Center do surveys, interviewing people. In hopes of getting an honest response, to find out their real needs, they send someone in anthropology or social sciences. "We have a student to do these interviews. I trust him; he's good, and this makes the results more valid." These data are combined with the facts about erosion, and weather patterns and rainfall have to be taken into account, plus the viability of solutions. Do they plant trees or wheat, restore wetlands or pave the place? The survey is designed to create a big picture. "Otherwise it is impossible to produce a report that will persuade decision-makers. Politicians must have an integrated view of the system, a summary of what the different experts are telling you. They have to delegate to get feedback. A politician may do that badly, or he may do it well. But they don't have much help; they're certainly unarmed in that sense. If you don't understand the system you're dealing with, what hope have you got?"

The system in question may be an entire river basin, but Allen, in his holistic view, says, "Any one perch is what it is because it sits where it does in a larger system. That's true of any fact."

He uses a computer program to set up a system with all the factors in it. This program is much more detailed and specific than the hologram, meant as a visual and symbolic model. Yet there are times when little clusters appear "out of nowhere," because the formulas for self-organization have been incorporated and reflect what can happen in life.

One project was a river polluted by fertilizers. The computer model for this includes a spatial representation of the river basin, the people whose lives are affected, the treatment plants, the wetlands. The context includes rainfall patterns, and the geology of the terrain, with slopes of runoff and vegetation. Numbers are entered in and graphs created. In the end, the computer program can be played with, changing parameters. The goal is to come up with a reasonable policy that benefits the environment, local agriculture, and does not cripple the chemical industries.

Playing with a parameter may mean posing a question: "We may ask farmers not to put so much nitrate on their land, and then see what happens," in the artificial computer model. "But first, we find out from farmers what they *would* do. We ask them. We put that into the model, then run the model to explore what might happen. Reducing a certain amount of nitrate may make a big difference, it may make no difference, it may put everyone out of business, or it may be necessary only in certain areas." The possibilities are figured at one year, five years, ten years, and so on. The market goes up, the market goes down. The rainfall fluctuates. "What will the chemical industries do if you say you can't do this this and this? Will they close down and go to Africa, which is what some of them are doing? We ask them. We try to be an interface between human decision systems and the ecology that they're relying on."

What looked like a good idea in the short term may not look so hot in time because change is one of the factors. Also, many development programs in Africa are notorious for imposing an outsider's notion of a solution and not asking the people how they would prefer to do it. "But the EEC is quite an intelligent organization," Allen says. "You've got this mix of cultures—if the English ever played their proper role, it would add a sense of pragmatism. You do have the French and the Italians who are real dreamers, Spanish, too; that whole Mediterranean block. They're into thinking. The science side of EEC is very good. They do research on the environment in a vast range of areas, and they have quite a lot of money. They saved European science, even in Germany, where funds were cut. EEC is trying to aim their science toward practical problems and potential policy issues. Desertification, river and water problems."

One river project was the Senegal, which runs on the northern boundary of the country. "It's about the only thing they've got left as a card," Allen says. "They built two dams and they're trying to develop paddy fields for rice. It's all pretty disastrous, but I think we can help them. If we build a model and introduce paddy fields at a certain rate, putting costs and so on into it, we can try to determine if they can produce rice at an affordable price." Some elements that he must fold into the model include the people who work there, the effect on migration patterns, and the market. Who will buy the rice? What is the international market? Bottomline: Is the terrace both economical and Green?

"It's very complicated. If you come in and say I want to take out the terraces, put in machinery and flatten this, because I can make more money, then you find you've got a massive erosion problem on your hands." Downstream other things may happen as a result. "There's all kinds of things linked."

In a North Sea project, Allen researched the effects of the fishing industry. "People assumed that the fish population was in equilibrium in nature. This seems absurd, because when you look at the figures they go up and down all the time. This is a very serious issue in terms of sustainable development, where people assume supply will meet demand. But sustainability is about resilience and response and adaptability."

Some factors that went into this computer program included: production rate, number of eggs, number of adults producing, and death rate from natural causes and from fishing. Then Allen added the human factors: number of boats, profits, fish market price changes. He drew from fishing records of the past twenty years.

At first, "the fishing industry felt things were in balance," Allen said. But due to oceanic fluctuations, "fish have a very uneven birth rate. Any new policy should be aiming at an uncertain resource, and the effects of the boom and bust. What I discovered [with the program] is that if you moved it to their idea of sustainability, you'd have a very large fleet of boats and very few fish at a very high price. That's what happened to lobsters; only the rich could afford them. So it depends on elasticity of demand. Management has to be about fish as a food supply and as a source of jobs."

For a project in Greece, orange farmers had drilled water holes for irrigation "the same way that fishermen catch fish," he said wryly. "They made a lot of money at first, but then the water table dropped 60 meters. Now the table is pulling in sea water. Rather than give them fines or restrictions, the first step was to understand what drove them to that, and try to solve the problem by solving their needs."

Does a model ever fail? "No, because we don't predict; our models are flexible, they're learning tools. You cannot have perfect knowledge. If the model doesn't agree with what happens, then you're learning something." But how do his clients feel about this? Allen implies that it may be imperfect, but it is the most realistic form developed thus far.

"We have to accept that uncertainty," he said, "instead of trying to develop methods simply because we want to be sure. There's no universal answer or recipe, there's no one way of doing things. If there is only one way, it isn't great. Pliability gives play to diversity, and keeps the strategies rich."

But can a computer model truly represent all the diversity that happens in life? "Modeling is the only thing we have; our thought processes use models, explicitly and implicitly. You don't make *the* model, and it finishes. The real world will go on changing. Your model will continue to be inadequate in various ways, which you will understand only if you have a model. Otherwise it's all just a mess."

So what he has done is create a blueprint, based on nature. The pattern in this case is life on earth. The formulas that Prigogine devised can be applied.

"Models are not as foreign as you think, because mathematics is only a language. Every equation can be said in words. For example, the change in the number of people in this area in the last six months is the number of people born, minus the number of people who died, plus the ones who came in, minus the ones that left. And that's an equation."

We had been talking for four hours, and Allen suggested lunch at a local British pub. On a winding road that cuts through cow pastures, I asked about the geology of the area. Bedford is the name of the local clay for the bricks that built London. Now the quarries are being filled with London's waste. Over shepherd's pies, we talked about recycling.

"If it takes a lot of energy to recycle something, then you haven't gained anything," Allen said. "We know this from entropy. If you have waste, your choices are dispersion or concentration."

"That we're all made of the same things, with an interchangeability of parts, is a lesson for technology. Chemicals could be recycled, but we produce compounds that are not biodegradable. It would be much better to design things so the elements can be disassembled and used again. We need to see an industrial metabolism, look at industry in biological ways. If you incinerate stuff and scatter it, it gets into the rain, into the oceans and rivers, and comes back anyway. Concentration of waste is better, because it not only concentrates the pollution but concentrates the mind." In other words, people will think about it if it has to be buried or transported in bulk, but simply burning it—out of sight, out of mind.

An American entrepreneur named Bill Haney has devised a solution for some waste. As a student at Harvard, Haney studied the history of science, which impressed on him the value of change in technology. In 1980, when he was 22, Haney borrowed $10,000 to establish Fuel Tech, Inc., and developed methods for breaking down pollution gases like the oxides of nitrogen. Seven years later, he sold Fuel Tech for $15 million, and then founded two other companies, Energy Biosystems and Molten Metal Technology, Inc. Molten Metal uses a heat bath to break down and extract waste compounds, converting them into raw materials that can be reused—metal alloys, ceramics, and gases. BioSystems plans to use microbes to remove sulfur from petroleum products, a process now requiring a great deal of energy at oil refineries. Haney's goal is to be the "leading developer of technology to handle waste streams."

Allen thinks the Eco-Center is unique because "very few people try to spawn an economic science policy." He adds, "On top of that, we have this framework of thinking which comes from the evolutionary system. This gives us an inside which nobody else has." Yet Bill Haney came up with a viable solution, based on his studies of the history of science and the logic of breaking down synthetic compounds. If you look at what happens within his Molten Metal bath, it resembles self-organization. The heat of the metal bath causes the waste compounds to break down and recombine to form new substances. The thermodynamic influence is powerful; temperatures of 3,000 degrees Fahrenheit can be used to reduce complex, hazardous waste, like PCBs, into hydrogen. It may also be applied to nuclear waste; under heat, the radioactive elements concentrate in a slag that forms on the surface. When it cools, it immobilizes the radioactive elements in a glassy solid.

The Sante Fe Institute applies physics to economics, and complexity is their focus. In May of 1994, the center in New Mexico sponsored a workshop called "Limits to Scientific Knowledge," sponsored by the Alfred P. Sloan Foundation in New York. Ralph E. Gomery, president of the foundation, is the former director of research for IBM, and he joined an interdisciplinary team to echo what Peter Allen had told me. No models can predict perfectly, and there is an "incompleteness" to math.

Kurt Gödel suggested this in the thirties, with the notion that many complex systems can never be proved or disproven, because the random interferes.

Allen says that the model of evolution is often ignored because "We have a serious problem in our mind in that what we mean by explanation implies determinism. Well, this happened before in the market, so it will happen again. There's a whole ecology of things to consider when you put things into an evolutionary perspective. This is also the resilience and the imagination in a system. We never know what's going to happen with certainty, yet innovation and metamorphosis exist.

"Take your book, for example. There are a million potentials. You will come to an end product through a series of accidents. It will be a success or a failure or in between through another series of accidents. And you can never put a button on why, or what the other potentials could have done. Life," he says, "is one of a million things all the time. I think this is too worrying for most people."

It wasn't too worrying for Darwin, who saw the greater flux on this earth spinning in space, and found "grandeur in this view of life," the order and flux of Mackendrick films, with two sides of a world that never come to rest. Darwin contrasted the steady course of the planet to the many changes taking place on earth, writing that while "this planet has gone cycling on according to the fixed laws of gravity, from so simple a beginning endless forms most beautiful and wonderful have been, and are being evolved."

D'Arcy Thompson managed to link these two worlds, showing how physical laws can influence life, and he had a grasp of the greater flux. He saw the "nonequilibrium" phase of cambium that Claus Mattheck uses to refine blueprints, the generating patterns that intrigue Dolf Seilacher, and the diffusion patterns that Prigogine would explore to discover dissipative structures. While D'Arcy Thompson did not confide any Buddhist inclinations, he did write, "We learn and learn, but never know all, about the smallest, humblest thing."

◆   ◆   ◆

ACKNOWLEDGMENTS

Aristotle appealed to the ego of every author when he suggested: "The greatest thing of all is to be a master of the metaphor. It is the only thing which cannot be taught by others; and it is also a sign of original genius, because a good metaphor implies the intuitive perception of similarity in dissimilar things." The genius that appears within these pages belongs to others.

Stephen Jay Gould, friend and mentor, cultivated my interest in "the peculiarities that inform, the generalities that instruct." I have drawn on several autobiographies, biographies and profiles of Buckminster Fuller including *An Autobiographical Monologue/Scenario; Bucky, A Guided Tour of Buckminster Fuller,* and *The Dymaxim Wold of Buckminster Fuller.* Much of the detail about the private life of D'Arcy Thompson relies on a biography, *D'Arcy Wentworth Thompson, The Scholar-Naturalist,* by Ruth Thompson, and insightful comments on his work by Gould, John Tyler Bonner, Sir Peter Medawar, and Clifford Dobell. There are references for sources by chapter and suggestions for further reading on page 216.

My mother (my high school instructor in geometry and algebra) showed me my first slide rule. Years later, by way of thanks, I brought her an abacus from Shanghai. I am grateful to the many people who endured interviews and shared their ideas and publications: Neill Alexander, Peter Allen, Martin Cowley, John Currey, Caroline Dry, J. E. Gordon, Michael Hausman, Mimi Koehl, Meave Leakey, J. William Littler, Paul MacCready, Claus Mattheck, Dolf Seilacher, Jeremy Rayner, Malcolm Telford, Steve Vogel, Alan Walker. Rebecca Finnell, associate editor at *Natural History*, and architect Craig Abel provided a deluge of books and papers, and exhibit designer Hal Chaffee and sculptor Fred Eversly provided tips on engineering. Mike Skrill loaned me his slide rules; Mary Anne Fitzgerald and Alyson Whyte helped with research in the U.K., and Paul Monsour helped with math.

I am grateful to John Brockman and Katinka Matson for being friends as well as agents, and at Addison-Wesley, Tiffany Cobb deserved first prize for patience had it not already been given to Bill Patrick, whose telephone calls had the effect of Valium.

Thanks to the organizers of SFB 230 and the Institute for the Study of Lightweight Structures, the Society of Experimental Biology, Leo W. Buss at Yale, Martin Cassidy at the American Museum of Natural History, Jill Lynne of Liaison Communications, friends Jane White, Raquel Buhrer, and Laura Utley of Global Communications for Conservation, Ian Wilson, Ken and Julie Slavin, Suzanne Klein and Hank Collins and their Dutchess County tree surgeons, who wore windbreakers that read "Rake and Roll."

A percentage of profits from this book will go to Global ReLeaf, a division of American Forests, established in 1875. The group plants trees in areas that have been damaged by erosion, fire, disease, and hurricanes. Some projects have included restoring 50 acres of ponderosa pine near Casper, Wyoming, in an area destroyed by fire; planting a half million trees in the area of south Florida ravaged by Hurricane Andrew; and restoring a tropical rain forest in the Hakalau National Wildlife Refuge in Hawaii. Ten dollars plants ten trees.

Global ReLeaf
American Forests
P.O. Box 2000
Washington, D.C. 20013
Tel. 1-800-873-5323

REFERENCES

Author's Notes

Tomlinson, P.B. "Tree Architecture," *American Scientist* 71 (1983): 141–149.

Hyman, Libbie. *The Invertebrates.* Vol. 4, *Echinodermata.* New York: McGraw-Hill, 1955.

Chapter One

de Saint-Exupéry, Antoine. *Wind, Sand and Stars.* New York: Harcourt Brace, 1940.

Gordon, J.E. *Structures, Or Why Things Don't Fall Down.* New York: Da Capo, 1978.

Wainwright, S.A., W.D. Biggs, J.D. Currey, and J.M. Gasline. *Mechanical Design in Organisms.* Princeton, NJ: Princeton University Press, 1982.

Petroski, Henry. *Beyond Engineering.* New York: St. Martin's, 1986.

Chapter Two

Fuller, Buckminster. *An Autobiographical Monologue/Scenario.* Edited by Richard Snyder. New York: St. Martin's, 1980.

Thompson, D'Arcy. *On Growth and Form.* London: Cambridge University Press, 1917, 1942. Abridged, edited by John Tyler Bonner. London: Cambridge University Press, 1961.

Mattheck, Claus. *Trees, The Mechanical Design.* New York: Springer-Verlag, 1991.

———. "Design and Growth Rules for Biological Structures and their Application to Engineering," *Fatigue of Engineering Materials* 13 (1990): 535–550.

———. "Why They Grow, How They Grow: The Mechanics of Trees," *Arboricultural Journal* 14 (1990): 1–17.

———. "A New Method of Structural Shape Optimization Based on Biological Growth," *International Journal of Fatigue* 12 (1990): 185–190.

———. "Soft Kill Option: The Biological Way to Find an Optimum Structure Topology," *International Journal of Fatigue* 14 (1992): 387–393.

———. "Successful Shape Optimization of a Pedicular Screw," *Medical and Biological Engineering and Computing* July (1992): 446–448.

———. Lecture, SFB 230, October 1991. "Tree design and tree failure: Computer simulation of growth and breakage," co-author Dagman Erb, Natural Structures conference, Stuttgart.

Jeronimidis, G. "Wood, One of Nature's Challenging Composites." In *Mechanical Properties of Biological Materials*, edited by John Currey and Julian Vincent.

Wainwright, Stephen A. *Axis and Circumference, The Cylindrical Shape of Plants and Animals*. Cambridge, MA: Harvard University Press, 1988.

Alvarez, A. "Sinking Fast; The World's Largest Ships Are Going Down in Record Numbers." *New York Times Magazine*, May 17, 1992.

Vogel, Steve. "When Leaves Save the Tree," *Natural History*, September 1993.

———. *Life's Devices*, Princeton University Press, 1988. As quoted in "An Engineer's Eye Helps Biologist Understand Nature," by Bruce Fellman, *Smithsonian*, July 1989.

National Register of Big Trees. Washington, D.C.: U.S. Forestry Association, 1992.

Wilson, B.F. *The Growing Trees*. Amherst, MA: University of Massachusetts Press, 1970.

Kozlowski, Theodore T. *Tree Growth and Environmental Stress*. Seattle: University of Washington Press, 1979.

Hoffman, Donald. *Frank Lloyd Wright, Architecture and Nature*. New York: Dover, 1986.

Angier, Natalie. "Warmings? Tree Rings Say Not Yet," *New York Times*, 1 December 1992.

Burdick, Alan. "Pie in the Sky?" *New York Times Magazine*, November 7, 1993.

McMahon, Thomas A., and John Tyler Bonner. *On Size and Life*. New York: Freeman, 1983.

LaBarbera, Michael. "Inner Currents," *The Sciences*, September/October 1991.

Chapter Three

Gosnell, Mariana. *Zero, Three, Bravo*. New York: Knopf, 1993.

McLanathan, Richard. *Leonardo da Vinci*. Notes for IBM exhibit.

McPhee, John. *The Deltoid Pumpkin Seed*. New York: Farrar, Straus & Giroux, 1973.

Dobbs, Edwin. "Paul MacCready and His Marvelous Machines." *Reader's Digest*, August (1991): 62–67.

Parrish, Michael. "The Big Guy's Back . . . Creating a Flap." *Smithsonian*, March 1986.

Dalton, Stephen. *The Miracle of Flight*. New York: McGraw-Hill, 1977.

Shlain, Leonard. *Art and Physics*. New York: Morrow, 1991.

Adams, Rick A., and Scott C. Pedersen. "Wings on Their Fingers," *Natural History*, January 1994.

Alexander, R. McNeill. *Exploring Biomechanics*. New York: Freeman, 1992.

Langston, Wann Jr. "Pterosaurs," *Scientific American* 244 (1981): 122–136.

Rayner, Jeremy M.V., G. Jones, and A. Thomas. "Vortex Flow Visualizations. . . ," *Nature* 321 (1986): 162–164.

Rayner, Jeremy M.V. "Mechanics and Physiology of Flight in Fossil Vertebrates," *Transactions of the Royal Society of Edinburgh, Earth Sciences* 80 (1989): 311–320.

———. "The Evolution of Vertebrate Flight," *Biological Journal of the Linnean Society* 34 (1988): 269–287.

———. "Pleuston: Animals Which Move in Water and Air." Vol. 10, *Endeavor*, Pergamon Journals, 1986.

Welty, Carl. "Birds as Flying Machines," *Animal Engineering*, San Francisco: Freeman, 1955.

Padian, Kevin. "The Origins and Aerodynamics of Flight in Extinct Vertebrates," *Palaeontology Review* 28, Part 3 (1985): 413–433.

Chapter Four

Gleick, James. *Chaos*. New York: Penguin, 1987.

Thompson, Ruth. *D'Arcy Wentworth Thompson, The Scholar-Naturalist*, Postscript by P.B. Medawar. Oxford: Oxford University Press, 1958.

von Frisch, Karl. *Animal Architecture.* New York: Harcourt Brace Jovanovich, 1974.

Le Gros Clark, W.E., and P.B. Medawar, eds. *Essays on Growth and Form.* London: Clarendon, 1945.

Kenner, Hugh. *Bucky, A Guided Tour of Buckminster Fuller.* New York: Morrow, 1973.

Nelson, George. Introduction, MAN transFORMS, Smithsonian Institution, Washington, 1976.

Fuller, Buckminster. "Synergetics," MAN transFORMS, Smithsonian Institution, Washington, 1976.

Dobell, Clifford. Obituary of D'Arcy Thompson, *Royal Society* 6 (1949): 598–617.

Gould, Stephen Jay. "D'Arcy Thompson and the Science of Form," *New Literary History*, vol. 2 (2) (1971): 229–258.

Hutchinson, G. Evelyn. "Marginalia," *American Scientist* 36 (1948): 577–606.

Thompson, D'Arcy. *Science and the Classics.* Oxford: Clarendon, 1940.

———. "Fifty Years Ago at the Royal Society." Address to the Royal Society of Edinburgh, 1934.

———. "Morphology and Mathematics," Trans. *Royal Society of Edinburgh* vol. L, Part 4, no. 27 (1915).

Thompson, D'Arcy. *On Growth and Form.* London: Cambridge University Press, 1917, 1942. Abridged, edited by John Tyler Bonner. London: Cambridge University Press, 1961.

———. *On Aristotle as a Biologist, With a Prooemion on Herbert Spencer,* Oxford: Clarendon, 1913.

Stiles, Janet. "Complexity." *Omni,* May 1994.

Chapter Five

Barnes, Julian. "The Maggie Years," *New Yorker,* January 6, 1992.

McPhee, John. "The Curve of Binding Energy." In *The John McPhee Reader.* New York: Farrar, Straus and Giroux, 1976.

Paulos, John Allen. *Innumeracy.* New York: Hill & Wang, 1989.

Thompson, D'Arcy. "On Leaf Arrangement or Phyllotaxis." In *On Growth and Form.* London: Cambridge University Press, 1942.

———. "Aristotle the Naturalist." *Science and the Classics.* In London: Clarendon, 1940.

Fletcher, Rachel. *Harmony by Design.* Chicago: Beverly Russell, 1993.

Grillo, Paul Jacques. *Form, Function and Design.* New York: Dover, 1975.

Ray, C. Claiborne. "Q & A Science," *New York Times,* 10 May 1994.

Thompson, Ruth. *D'Arcy Wentworth Thompson, The Scholar-Naturalist.* Postscript by P.B. Medawar. Oxford: Oxford University Press, 1958.

Ghyka, Matila. *The Geometry of Art and Life.* New York: Dover, 1977.

Wainwright, Stephen A. *Axis and Circumference, The Cylindrical Shape of Plants and Animals.* Cambridge, MA: Harvard University Press, 1988.

Littler, J. William. "On the Adaptability of Man's Hand," *The Hand* 5 (1973): 187–191.

Gould, Stephen Jay. *The Panda's Thumb.* New York: Norton, 1980.

Gould, S.J., and R.C. Lewontin, "The Spandrels of San Marco and the Panglossian Paradigm: A Critique of the Adaptationist Programme," *Proceedings of the Royal Society, London [B]* 205, (1979): 581–598.

Hambidge, Jay. "The Diagonal." 1919. Reprinted in *The Elements of Dynamic Symmetry.* New York: Dover, 1967.

Lopez, Barry. *Arctic Dreams.* New York: Scribner, 1986.

Bonner, John T., and Tom McMahon. *On Size and Life.* Scientific American Books, New York: Freeman, 1983.

Moore, Randy. "The Numbers of Life," *American Biology Teacher* 54 (1992): 68.

Ferris, Timothy. *Coming of Age in the Milky Way.* New York: Morrow, 1988.

Broad, William J. "Top Quark, Last Piece in Puzzle of Matter, Appears to Be in Place," *New York Times,* 26 April 1994.

Teresi, Dick. "Perhaps This Universe Is Only a Test." *New York Times Book Review,* September 5, 1993.

Weiss, Rick. "Techy to Trendy, New Products Hum DNA's Tune," *New York Times,* 8 September 1992.

Gordon, J.E. *Structures, Or Why Things Don't Fall Down.* New York: Da Capo, 1978.

Chapter Six

Marks, Robert W. *The Dymaxion World of Buckminster Fuller.* New York: Reinhold, 1960.

Fuller, Buckminster. *An Autobiographical Monologue/Scenario.* Edited by Richard Snyder. New York: St. Martin's, 1980.

Kenner, Hugh. *Bucky, A Guided Tour of Buckminster Fuller.* New York: Morrow, 1973.

Updike, John. "Things, Things." *New Yorker,* January 18, 1993, 104–106.

Hoffman, Donald. *Frank Lloyd Wright, Architecture and Nature.* New York: Dover, 1986.

Koenig, Rhoda. "A Spanish Disquisition." *New York,* February 10, 1992.

Martinell, Cesar. *Gaudi, His Life, His Theories, His Work.* Edited by G.R. Collins. Cambridge, MA: MIT Press, 1975.

Read, Mimi. "A Wright Disciple Now Rivals the Master Himself," *New York Times,* 10 October 1991.

Drew, Philip. *Frei Otto: Form and Structure.* Boulder, CO: Westview, 1976. Includes the essay "The Lightweight Aesthetic."

Fuller, R. Buckminster. *R. Buckminster Fuller on Education,* a series of lectures. Edited by Peter H. Wagschal and Robert D. Kahn. Amherst, MA: University of Massachusetts Press, 1979.

Thompson, D'Arcy. *On Growth and Form.* London: Cambridge University Press, 1917, 1942. Abridged, edited by John Tyler Bonner. London: Cambridge University Press, 1961.

Rothman, Tony. "Geodesics, Domes and Spacetime." In *Science à la Mode.* Princeton, NJ: Princeton University Press, 1989.

Otto, Frei. "Animate Structures and Technical Structures," and, with J.G. Helmcke, "Lebende und technische Konstruktionen." In *Deutsche Bauzeitung* 67 (1962): 855–861.

Gordon, J.E. *Structures, Or Why Things Don't Fall Down.* New York: Da Capo, 1978.

Chapter Seven

Gould, Stephen Jay. *Wonderful Life, The Burgess Shale and the Nature of History.* New York: Norton, 1989.

Attenborough, David. *Life on Earth.* Boston: Little Brown, 1980.

Reader, John. *The Rise of Life, The First 3.5 Billion Years.* New York: Knopf, 1986.

Symposium Honoring Professor Willard Hartman. The Peabody Museum of Natural History, Yale University, New Haven, CT, May 5, 1992.

Seilacher, Adolf. "Distinctive Features of Sandy Tempestites." In *Cyclic and*

Event Stratification, edited by Einsele and Seilacher. New York: Springer-Verlag, 1982.

————. "Constructional Morphology of Sand Dollars," *Paleobiology* 5 (1979): 191–221.

————. "Self-Organizing Mechanisms in Morphogenesis and Evolution." In *Construction Morphology and Evolution*, edited by Norbert Schmidt Kittler and Klaus Vogel. New York: Springer-Verlag, 1991.

Thompson, D'Arcy. "On the Shapes of Eggs and of Sea Urchins," in *On Growth and Form*. London: Cambridge University Press, 1917.

Raup, David M., and Adolf Seilacher. "Fossil Foraging Behavior: Computer Simulation," *Science*, 166 (1969): 994–995.

Koehl, M.A.R. "The Interaction of Moving Water and Sessile Organisms," *Scientific American*, December 1982.

"MacArthur Grants." *California Monthly*, September 1990.

Dean, Cornelia. "At Edge of Beach, 2 Theories at Odds," *New York Times*, 10 August 1993.

Chapter Eight

Chartrand, Sabra. "Patents," *New York Times*, 7 June 1993.

Hildebrand, Milton. "The Epsom Derby," *BioScience*, December 1989.

Hagen, Charles. "Eadweard Muybridge Revisited, With Others," *New York Times*, 6 March 1992.

Dalton, Stephen. *The Miracle of Flight*. New York: McGraw-Hill, 1977.

————. *Borne on the Wind*. Reader's Digest Press, 1975.

Braun, Marta. *The Work of Etienne-Jules Marey*. Chicago: University of Chicago Press, 1993.

Woakes, A.J., and W.A. Foster. "The Comparative Physiology of Exercise." *SEB Journal*, The Company of Biologists, Ltd., Birmingham, April 1991.

Alexander, R. McNeill. "Biomechanics in the Days Before Newton," *New Scientist*, September 30, 1989.

————. *Exploring Biomechanics*. New York: Freeman, 1992.

————. *Locomotion of Animals*. New York: Chapman & Hall, 1982.

————. *Optima for Animals*. London: Arnold, 1982.

————. "The Spring in Your Step." Lecture to Royal Institution, London, 1992.

Jerome, John. "Profile of Tom McMahon." *Harvard*, November/December 1991.

Kunzig, Robert. "The Man Who Weighs Dinosaurs." *Discover*, October 1990.

Rayner, Jeremy. "The Cost of Being a Bat," *Nature* 350 (1990).

Diamond, Jared M. "How to Fuel A Hummingbird," *Nature* 348 (1990).

Wells, Martin. "The Dilemma of the Jet Set." *New Scientist*, February 17, 1990.

Dry, Caroline. "Passive Smart Materials for Sensing and Actuation," University of Illinois at Champaign, Architecture Research Center. Abstract in 1992.

Begley, Sharon, with Carolyn Friday, "Nature at the Patent Office," *Newsweek*, December 14, 1992.

Coghlan, Andy. "Smart Ways to Treat Materials." *New Scientist*, July 4, 1992.

Browne, Malcom W. "Pianists Tickle the Polymers," *New York Times*, 25 May 1993.

Amato, Ivan. *Science* 255 (1992): 284–286.

Siwolop, Sana. "A Sticky Wicket: Conjuring Up Formulas for Sealants," *New York Times*, 3 January 1993.

"How Smart Is Concrete?" *Engineering News Review*, March 23, 1992.

Hazleton, Lesley. "Really Cool Cars." *New York Times Magazine*, March 29, 1992.

"GM Putting Electric Car to Test," *New York Times*, 28 January 1994.

Chapter Nine

Kenner, Hugh. *Bucky, A Guided Tour of Buckminster Fuller.* New York: Morrow, 1973.

Holusha, John, and Bill Haney. "From Firewood to Environmental Empire in 14 Years," *New York Times*, 26 June 1994.

Darwin, Charles. *On the Origin of Species.* London: John Murray, 1859.

Mosbrugger, Volker. *The Tree Habit of Land Plants, A Functional Comparison of Constructions with a Brief Review on the Biomechanics of Trees*, SFB 230, 1988.

INDEX

224

PERMISSION ACKNOWLEDGMENTS

"The Legend of the Sand Dollar" on page xix reprinted with permission of Aerial Photography Services, Inc., Charlotte, N.C.

Portions of chapter one first appeared in *Audubon* magazine.

Quotation on page 1 from *Mr. Noon,* by D.H. Lawrence. Reprinted with permission of Laurence Pollinger Ltd. and the Estate of Frieda Lawrence Ravagli.

Quotation by D'Arcy Thompson on page 15 from *On Growth and Form* reprinted with permission of Cambridge University Press.

Drawings on pages 39 and 70 from *On Growth and Form* by D'Arcy Thompson. Reprinted with permission of Cambridge University Press.

Quotation on page 40 from *Mating* by Norman Rush. Reprinted with permission of Random House, Inc.

Quotation by Paul MacCready on page 40 from *Zero, Three, Bravo* by Mariane Gosnell reprinted with permission of Alfred A. Knopf, Inc.

Photographs on page 46 from "Biomechanical Aspects of the Wing Joints in Flies," by A. Wisser and W. Nachtigall in *Constructional Morphology and Evolution*. Reprinted with permission of Springer-Verlag New York, Inc.

Quotation on page 92 from "The Maggie Years," by Julian Barnes, from the November 15, 1993 issue of *The New Yorker*. Reprinted with permission of *The New Yorker*.

Illustrations on pages 96 and 97 by Chris St. Cyr.

Illustration on page 103 reprinted with permission of J. William Littler, M.D.

Photograph on page 126 reprinted with permission of Ampliciones y Reproducciones Mas (Arxiu Mas), Barcelona, Spain.

Photograph on page 133 reprinted with permission of Buckminster Fuller Institute, Santa Barbara. Copyright 1960 Allegra Fuller Snyder.

Photograph on page 145 reprinted with permission of the Institute for the Study of Lightweight Structures, Stuttgart.

Quotation on page 149 from *Wonderful Life* by Stephen Jay Gould. Reprinted with permission of W.W. Norton & Company.

Illustrations on pages 152, 164, and 167 reprinted with permission of Adolf Seilacher.

This book contains quoted material from *D'Arcy Wentworth Thompson, The Scholar-Naturalist* by Ruth Thompson. Copyright 1958, reprinted with permission of Oxford University Press.

This book contains quoted material from *On Growth and Form* by D'Arcy Thompson. Reprinted with permission of Cambridge University Press.

Photographs on pages xx, xxii, 18, 19, and 94 are by the author.